Commercial Aircraft Composite Technology

Ulf Paul Breuer

Commercial Aircraft
Composite Technology

Springer

Ulf Paul Breuer
Institut für Verbundwerkstoffe GmbH
Kaiserslautern, Germany

ISBN 978-3-319-81153-6   ISBN 978-3-319-31918-6   (eBook)
DOI 10.1007/978-3-319-31918-6

Printed on acid-free paper

This Springer imprint is published by Springer Nature
The registered company is Springer International Publishing AG Switzerland

Preface

The contents of this book are based on my lectures "Commercial Aircraft Composite Technology" (*Verbundwerkstoffe im Flugzeugbau*) given to master's students at the Faculty of Mechanical Engineering, Technical University of Kaiserslautern, Germany, who wish to broaden their specific knowledge about composite material properties and manufacturing-optimised light weight design. The topic is very extensive and comprehensive textbooks only exist for certain areas. I have tried to focus on the central theme of societies' overall aircraft requirements to specific material requirements and to highlight the most important advantages and challenges of carbon fibre-reinforced plastics (CFRP) compared to conventional materials. During the product development process, it is fundamental to decide on "the right material at the right place" early on. It is, therefore, basic to understand the main activities and milestones of the development and certification process and the systematic defining of clear requirements. The process of material qualification—verifying material requirements—is explained in detail. All state-of-the-art manufacturing technologies are described, including future perspectives. Especially for composites it is key to understand the interaction of the design scheme, manufacturing technology and resulting material properties. Here I have tried to highlight some key aspects of advanced CFRP design for primary load carrying airframe structures along with selected examples. I have also included a short chapter on testing as part of the certification process—and repair.

As more and more high-performance composites such as CFRP are also used in other sectors—especially automotive, manufacturing systems engineering, wind power, architecture, sports and leisure and medical engineering—the textbook can also be useful for students and engineers engaged in these areas.

The book closes with an outlook on some of the latest developments for future aircraft that we presently pursue at the Institut für Verbundwerkstoffe (IVW). This last chapter was supported by my assistant researchers Tim Krooß (PPS—PES blends, enabling a material performance close to PEEK at lower cost), Benedikt Hannemann (carbon metal fibre hybrid materials, enabling electrical conductivity and improved failure behaviour) and Moritz Hübler (morphing structures by shape memory alloy integration into fibre-reinforced polymer structures).

I would like to thank all my old colleagues and friends for their thorough proofreading and their precious hints for improvements to this text.

Kaiserslautern, Germany   Ulf Breuer
Spring 2016

Contents

Abbreviations

A/B	Air brakes
A/C	Aircraft
acc.	According
ACD	Aircraft certification documents
ADL	Allowable damage limit
A_F	Austenite finish temperature
AFP	Automated fibre placement
Aft	Rear, back
AITM	Airbus industries test method
Al	Aluminium
ALI	Airworthiness limitation items
AMC	Acceptable means of compliance
AMM	Aircraft maintenance manual
approx.	Approximately
Ar	Argon
A_S	Austenite start temperature
ATA	Air Transport Association
ATL	Automated tape laying
Au	Aurum (gold)
bbl	Barrel of oil (359 l)
BBL	Body buttock line (fuselage width-wise position)
BS	Body station (fuselage lengthwise position)
BVID	Barely visible impact damage
BWL	Body water line (fuselage height-wise position)
C	Carbon
CAD	Computer-aided design
CAGR	Compound annual growth rate
CAI	Compression after impact
CB	Cross beam
CF	Carbon fibre

CF/EP	Carbon fibre/epoxy
CFRP	Carbon fibre-reinforced plastic
CMR	Certification maintenance requirements
Conf.	Configuration
cont.	Continued
cpt	Cured ply thickness
CRC	Corporate research centre
CRI	Certification review items
CTE	Coefficient of thermal expansion
Cu	Copper
CWB	Centre wing box
DEA	Dielectrical analysis
DMTA	Dynamic mechanical thermal analysis
DOC	Direct operating cost
DSC	Differential scanning calorimetry
DSG	Design service goal
EADS	European Aeronautic Defense and Space Company
EAS	Equivalent air speed
EASA	European Aviation Safety Agency
EIS	Entry into service
EKDF	Environment knock-down factor
EP	Epoxy
ERF	Electrorheological fluid
ESG	Extended service goal
ESN	Electrical structural network
F/S	Front spar
FAA	Federal Aviation Administration
FAL	Final assembly line
FAR	Federal aviation regulations
FAW	Fibre areal weight
FC	Flight cycle
FE	Finite element
FEA	Finite element analysis
FEM	Finite element method
FH	Flight hour
FHC	Filled hole compression (strength)
FHT	Filled hole tension
FML	Fibre metal laminate
FRP	Fibre-reinforced plastic
FST	Fire (or flammability) smoke toxicity
FSW	Friction stir welding
FT, F/T	Flap track
FTF	Flap track faring
FTI	Flight test installation

FTIR	Fourier transform infrared spectroscopy
FWD	Forward
GF	Glass fibre
GFRP	Glass fibre-reinforced plastic
GLARE	Glass laminate aluminium-reinforced epoxy
GSE	Ground support equipment
H/S	High speed
H/W	Hot/wet (condition)
HT	High tensile strength
HFEC	High frequency eddy current
HM	High modulus
HMS	High modulus/high strength
HST	High failure strain
HTP	Horizontal tail plane
ICAO	International Civil Aviation Organization
IE	Impulse echo
ILSS	Interlaminar shear strength
IM	Intermediate modulus
IPS	Individual product specification
IR	Infrared
ISO	International Organization of Standardization
IVW	Institut für Verbundwerkstoffe
JAA	Joint aviation authorities
JAR	Joint aviation requirements
JIT	Just in time
L	Longitudinal
L/S	Low speed
LBW	Laser beam welding
LDC	Large damage tolerance
LEF	Load elevation factor
Li	Lithium
LL	Limit load
LM	Low modulus
LSP	Lightning strike protection
LT	Longitudinal transversal
MAC	Mean aerodynamic chord
MCFRP	Metal-carbon-fibre-reinforced polymer
MCS	Multifunctional control surfaces
MEK	Methyl ethyl ketone (cleaning agent)
MF	Metal fibre
Mg	Magnesium
MG	Master geometry
Mini-TED	Mini-trailing edge device
MLG	Main landing gear

MoC	Means of compliance
MP	Maintenance programme
MPD	Maintenance planning document
MRO	Maintenance, repair and overhaul
MS	Material specification
MTOW	Maximum take-off weight
MVI	Modified vacuum infusion process
MWE	Manufacturing weight empty
MWNT	Multi-walled nanotubes
MZFW	Maximum zero fuel weight
NC	Numerical control
NCF	Non-crimp fabric
NDT	Non-destructive testing
NGA	Next generation aircraft
NLG	Nose landing gear
NRC	Non-recurring cost
OC	Operating COST
OHC	Open hole compression
OHT	Open hole tension
PAN	Polyacrylnitrile
Pax	Passengers
PB	Plumbum (lead)
PEEK	Polyetheretherketon
PES	Polyethersulfone
PLB	Pin loaded bearing (strength)
PLB ult.	Pin loaded bearing ultimate (strength)
PLC	Plain loaded compression
PLT	Plain loaded tension
PPS	Polyphenylene sulfide
ProHMS	Prozesskette Hochauftrieb mit multifunktionalen Steuerflächen
PZT	Lead zirconate titanate
QA	Quality assurance
R curve	Residual strength curve
R&D	Research and development
R/S	Rear spar
RC	Recurring cost
RF	Reserve factor
RFI	Resin film infusion
RI	Resin infusion
RI	Repeated interval
RPB	Rear pressure bulkhead
RPK	Revenue passenger kilometres
RT	Room temperature
RTM	Resin transfer moulding

RTM6	Special HexFlow® Hexcel Resin (Epoxy) for resin transfer moulding
SAM	Space allocation model
Sc	Scandium
SEM	Scanning electron microscopy
SF	Safety factor
SFRP	Steel fibre-reinforced plastic
SHM	Structural health monitoring
Si	Silicium
SMA	Shape memory alloy
SRM	Structural repair manual
SW	Sandwich
T	Transversal
T/C	Ratio air foil thickness to chord ratio
T/E	Trailing edge
tbd	To be determined (or: to be defined)
TC	Type certification
TEM	Scanning tunnelling electron microscopy
tex	Unit of measure for the mass per length of rovings or yarns, 1 tex = 1 g/1000 m
Ti	Titanium
T-L	Transversal-longitudinal
TP	Thermoplastic
TS	Technical specification
TTU	Through transmission ultrasound
UAV	Unmanned air vehicle
UD	Unidirectional
UL	Ultimate load
UN	United Nations
US	United States
UV	Ultraviolet radiation
V&V	Verification and validation
VAP®	Vacuum assisted process
VDI	Verein Deutscher Ingenieure
VID	Visible impact damage
Vol.	Volume
VSDI	Visual special detailed inspection
VTP	Vertical tail plane
XPS	X-ray photoelectron spectroscopy
Zn	Zinc

Symbols

A	Area
a	Crack length
b_F	Specific fuel burn
c_D	Coefficient of drag
c_L	Coefficient of lift
d	Diameter
da/dN	Crack growth rate
ε	Strain
E	Elastic modulus
F, f	Force
F_{tu}	Force tensile ultimate
F_{ty}	Force tensile yield
G	Gravity force
g	Earth acceleration
G_{ic}	Fracture toughness
k	Stress intensity factor
κ	Specific electrical conductivity
L	Lift force
M	Momentum
m	Mass
m_F	Fibre mass content
N, n	Number
n_Z	Load factor (number to multiply with g, earth acceleration)
p	Pressure
Δp	Delta pressure
R	Range
R	Residual strength
S	Electrical conductivity
ρ	Density
σ	Stress

τ	Shear stress
T	Temperature
t	Thickness
T_g, T_G	Glass transition temperature
V	Volume
v	Velocity
v_c	Velocity during cruise
v_D	Diving velocity
v_F	Fibre volume fraction
v_f	Velocity flaps
v_{fe}	Velocity flaps fully extended
w	Width

Chapter 1
Introduction

Abstract The world population is growing and the trend to further urbanisation is undamped. In addition, the economic development proceeds especially in the emerging countries. Hence, a further growth of the world air traffic can be expected, and the total number of commercial aircraft is expected to double within the next 15 years. At the same time, a large number of old aircraft must be replaced. Governments have set ambitious targets with respect to highest safety and highest environmental standards at even lower travel cost. New technologies are required to tackle these challenges by the next generation of aircraft. Composites offer additional product value and have increasingly been used for different parts of the airframe for almost 100 years of aviation history. After a discussion of the market development and a summary of the history and milestones of composites used for airframe structures, the true value of light weight, the most important material requirements and the key contributors as well as the main lever arms for even lighter future airframe are introduced.

Keywords Aircraft market • Future aircraft requirements • Advanced airframe materials • Next generation aircraft • Aircraft operational cost • Oil price development • Aircraft weight breakdown • Breguet equation • Value of light weight • Composite airframe history • Carbon fibre market • Life cycle analysis • CFRP recycling

General Airframe Material Requirements: The Value of Light Weight

Ever since the very first days of commercial air traffic at the beginning of the last century, engineers had to tackle the difficult challenge of providing a very low mass but high load carrying airframe structure, which had to fulfil several functions for a safe and fast transportation of passengers and freight across long distances:

- high load carrying capability at minimum weight
- passenger and pay load protection
- media resistance
- damage tolerance
- long service life

Fig. 1.1 Balance of forces

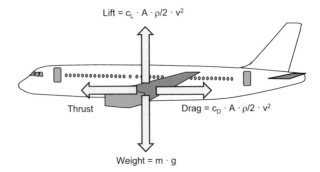

- manufacturability
- reparability
- affordability

This list is in no way exhaustive, but it is evident at first glance that the choice of materials had and still has a major impact on the fulfilment of these functions. Figure 1.1 illustrates the balance of forces of an aircraft during cruise in a simplified manner. Lift L is necessary to counteract the gravity force G, Eqs. (1.1) and (1.2), where m is the total aircraft mass and g earth gravity acceleration. The lifting force according to Eq. (1.3) is mainly created by the airflow around the aerofoils (wing Area A, coefficient of lift c_L, ρ is the air density), for which a relative movement (velocity v) of air and aircraft is needed.

However, this velocity with its square also causes drag, Eq. (1.4), where c_D is the coefficient of drag. The optimisation of the ratio c_D/c_L (i.e. trying to minimise drag during cruise) remains a difficult challenge of advanced aerodynamic design. For an A320, this ratio is approx. 1:20, for advanced gliders approx. 1:60. Thrust created by fuel consuming engines is necessary to counteract the drag and to maintain the necessary lift at a given cruise velocity. Higher masses require higher lift, thus producing higher drag, which results in higher thrust and a higher energy consumption.

Any mass reduction of airframe structure (or on board systems) can be directly converted into reduced fuel burn (less cost, lower emissions), higher payload (more passengers, more freight) or range extension (longer distances). Equation (1.5) is the Louis Charles Breguet equation of range, where v is the velocity, b_F is the specific fuel burn, c_D/c_L the drag/lift ratio, g the gravity acceleration, m_0 the mass of the aircraft at the beginning of the mission and m_t the mass of the fuel burnt during the mission. A typical value for a 1 kg weight saving of an A320-type aircraft is the equivalent of a 2000 l fuel saving over the complete life cycle of the aircraft (i.e. 2000 l is the sum of fuel savings of all its flights during its complete life time), which means a cost saving for the aircraft operator of around $1000, depending on the fuel price.

$$L = G \tag{1.1}$$

$$G = m*g \tag{1.2}$$

$$L = c_L*A*\frac{\rho}{2}*v^2 \tag{1.3}$$

$$D = c_D*A*\frac{\rho}{2}*v^2 \tag{1.4}$$

$$R = \frac{v}{b_F*\frac{c_D}{c_L}*g*ln\left(\frac{m_0}{m_0-m_t}\right)} \tag{1.5}$$

where
  L is the lift force [N]
  G is the gravity force [N]
  m is the mass of the aircraft [kg]
  m_0 is the mass of the aircraft at the beginning of the mission [kg]
  m_t is the mass of the fuel burnt during the mission [kg]
  g is the acceleration of earth [m/s^2]
  A is the wing area [m^2]
  ρ is the air density [kg/m^3]
  c_L is the coefficient of lift [1]
  c_D is the coefficient of drag [1]
  v is the velocity [m/s]
  R is the Range [m]
  b_F is the specific fuel burn [kg/N·s]

Aircraft Market Development

As more and more passengers are using aircraft for their transportation, the global economic impact of air traffic is impressive. According to [1], about 60 million jobs worldwide are directly linked to aviation, with a turnover of more than US$2.2 trillion.

However, the ecological impact is also impressive, and man-made CO_2 emissions from aviation are no longer negligible. With 3,300,000,000 passengers and 38 million tonnes of freight carried in 2014, the share of global CO_2 emissions was approx. 2 %, [1]. It is important to point out that in addition to the monetary value, there is also an ecological value of lightweight materials and the resulting fuel savings.

Figure 1.2 illustrates the typical situation of the airspace during the evening hours of a working day. With more than 20,000 registered aircraft in service in the year 2014 and more than one aircraft taking off or landing each second, air traffic is still growing, Fig. 1.3.

Fig. 1.2 American (*left*) and European (*right*) Airspace, on a typical working day at 20:30 (CET)
Courtesy of planefinder.net

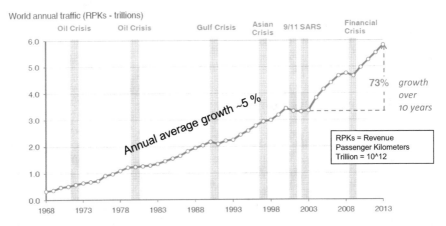

Fig. 1.3 Air traffic growth [2]

With an average growth of about 5 % per year, it can be expected that the number of passenger kilometres will double again within the next 15 years.

Cargo transportation is a growing market, too. According to [2], the annual growth is about 7 %, with more than 200 airlines and more than 1600 aircraft performing cargo only operations.

Main reasons are

• increasing world population
• urbanisation

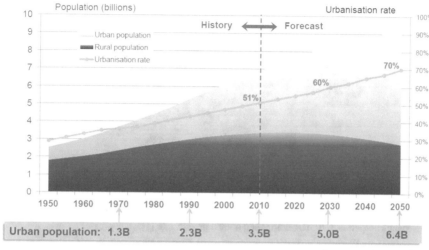

Fig. 1.4 World population development

- increasing wealth
- increasing efficiency of infrastructure
- decreasing air transportation cost

Figure 1.4 shows the world population development. It is expected that the share of urban population will increase within the next 40 years. This also means a higher availability of air transportation for a higher number of people. It is also expected that Asia-Pacific will play a more important role for air traffic growth than Europe and the U.S., Fig. 1.5.

Furthermore, emerging countries such as China, India, the Middle East, Asia, Commonwealth Independent States CIS (especially Russia), Africa, Latin America and Eastern Europe will play the most important role in terms of financially solvent middle class passengers; see Fig. 1.6.

The future development of aircraft size and capacities is a controversial issue. With respect to the share of all aircraft today, short to medium range aircraft with a typical passenger capacity of 100–250 seats and a design range of 2000–5000 nm are clearly dominating. According to [2], their share today (2015) accounts for more than 60 %.

Regional jets such as

- Antonov An-148
- AVIC ARJ-700
- Avro RJ70, RJ85
- BAe 146-100, -200
- Bombardier CRJ
- Dornier 328JET
- Embraer 170, 175

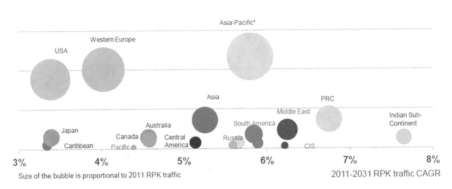

Fig. 1.5 Air traffic and traffic growth by regions, 2011 from/to/within traffic in RPK = revenue passenger kilometres, CAGR = compound annual growth rate [1]

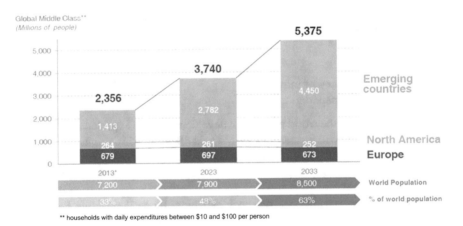

Fig. 1.6 Global middle class development [1]

- Embraer ERJ-135, -140, -145
- Fokker 70, F28
- Mitsubishi MRJ (EIS 2017)
- Sukhoi Superjet 100
- Yakovlev Yak-40

with typically less than 100 passengers contribute a market share of approx. 12 %.

Twin aisle and large jets for typically 200–400 passengers and a design range of up to 8000 nm have a market share of approx. 20 %. Small twin aisle aircraft are for example:

- Boeing 767, 787
- Boeing-MDC DC-10

- Airbus A300 (end of prod 2007)
- A310 (end of prod 1998)
- Airbus A330-200
- Airbus A330neo (EIS end 2017)*
- Airbus A350-800
- Lockheed L-1011
- Ilyushin IL-96

Medium twin aisle aircraft are for example:

- Boeing 777
- Boeing-MDC MD-11
- Airbus A330-300, A340
- Airbus A330neo (EIS end 2017)*
- Airbus A350-900, -1000
- Ilyushin IL-86

Only a small share of about 5 % can be attributed to very large aircraft with more than 400 passengers (A380, B747) and a design range of more than 8000 nm. Figure 1.7 illustrates cross sections of very large aircraft (A380, left), long range twin aisle aircraft (A340) and short to medium range single aisle aircraft (A320), right. Although according to [1] the number of world aviation centres such as Amsterdam, Los Angeles, Sydney etc. with more than 50,000 daily long-haul

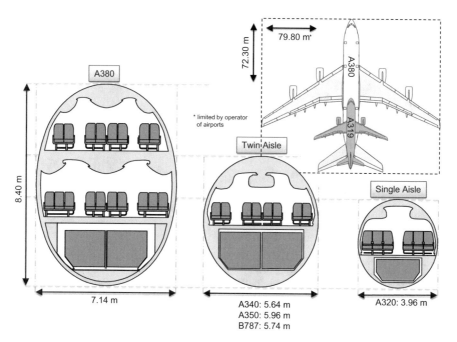

Fig. 1.7 Cross section and size comparison of small single aisle (A320), twin aisle (A340, A350, B787) and large (A380) aircraft, figures collected from [3]

passengers is expected to increase, it is not fully clear today if the market will require a higher number of very large aircraft in the future.

It is, however, undisputed that a very large number of short to medium range single aisle aircraft is needed, not only to cope with the rising number of passengers, but also to replace the present fleet of these types of aircraft. In [2] it is estimated that the total number of aircraft in service will rise from 20,910 in 2013 to 42,180 in the year 2033, of which more than 70 % would be short to medium range single aisle aircraft. As many of today's aircraft need replacement, Boeing estimates the need for 36,770 new aircraft until 2033, of which again 70 % would be short to medium range single aisle aircraft. Present airplanes of this type are for example:

- Boeing 717, 727
- Boeing 737-100 through -500
- Boeing 737-600, -700, -800
- Boeing 737 MAX 7, MAX 8
- Airbus A318, A319, A320
- Airbus A319neo, A320neo
- Boeing-MDC DC-9, MD-80, -90
- AVIC ARJ-900
- BAe 146-300, Avro RJ100
- Bombardier CRJ-1000
- Bombardier CS100, CS300
- COMAC C919 (EIS planned 2018)
- Embraer 190, 195
- Fokker 100
- Ilyushin IL-62
- Tupolev TU-154
- Yakovlev Yak-42
- UAC MS 21-200 -300

and with more than 175 seats:

- Boeing 707, 757
- Boeing 737-900ER
- Boeing 737 MAX 9
- Airbus A321
- Airbus A321neo
- Tupolev TU-204, TU-214
- UAC MS 21-400

This has a direct impact on production rates, production and assembly technology as well as on material choice. Today, the world production rate for this type of jets is at around 80 aircraft per month. Even if new manufacturers such as Comac in China will increase their production share in the future, it can be expected that more efficient manufacturing technology combined with appropriate materials will be a key factor to cope with the increasing demand for higher production rates and lower manufacturing cost for more affordable airframe structures. This will also promote automation, and manual manufacturing operations will be more and more reduced.

According to data available from electronically published online catalogues, catalogue prices range between 70 million euros for a small to medium sized single aisle aircraft and more than 400 million euros for a very large long range aircraft (A380). The catalogue price for the "all composite airframe" aircraft B787 is well above 200 million euros. However, in addition to the cost of capital needed by airline operators to acquire new aircraft, it is important to understand the main contributors of operational cost, such as

- ticketing
- flight crew wages
- fuel and oil expenses
- maintenance
- landing charges
- navigation fees
- station expenses
- passenger services
- insurance
- training
- depreciation and amortisation
- general administration and other

Figure 1.8 shows an operational cost breakdown of an A320-type aircraft [4] for 2003 and 2006 for maintenance, navigation, crew and fuel cost. The fuel cost portion clearly depends on the fuel price. It is, however, evident that the choice of airframe material and its design will directly affect a large portion of the cost, since not only fuel cost can be cut by more advanced lightweight design (refer to Eq. 1.5), but also maintenance cost. In case of "no crack growth" design concepts, as usually applied for CFRP airframe structures, inspection effort can be substantially reduced compared to conventional aluminium design solutions. The latter require careful and time consuming analysis of crack lengths, performed by

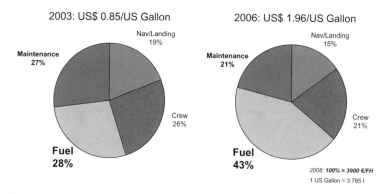

Fig. 1.8 Operational cost breakdown (A320) [4]

specially qualified personnel, in order to insure that cracks will not become critical within the scheduled inspection interval.

Next Generation Aircraft Requirements

The requirements for the next generation of aircraft defined by society and governments are extremely challenging. The "Vision 2020" [5] as well as "Flightpath 2050" [6] are defining important future scenarios, targets and recommendations such as:

- lower travel cost
- better service quality
- highest safety
- highest environmental standards

The targets can be structured by a "4C" principle: Cost, Comfort, Convenience, Choice. Cost: More efficient airline systems and reduced cost of ownership (with fuel savings, reduced maintenance effort etc.) shall enable operators to pass savings on to passengers. Comfort: More passenger-friendliness, less noise and vibrations and less disturbance by turbulences are expected, contributing to an "executive car feeling" comfort. Convenience: Airline schedules should become more reliable (99 % of flights on time as scheduled), more independence from weather, with short passages for boarding and de-boarding, and with information on demand for all passengers. Choice: More routes and more flights shall be made available in order to cope with the high number of passengers, and more customisation is expected, too. Some important environmental targets are depicted in Fig. 1.9. Compared to a year 2000 reference, "Vision 2020" claims for -50% noise footprint, -50% CO_2

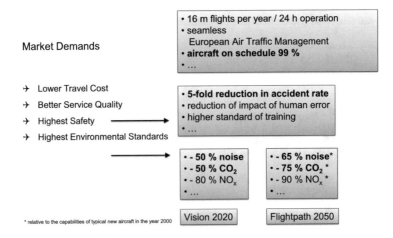

Fig. 1.9 "Vision 2020" and "Flightpath 2050" requirements

and -80% NO_x, while "Flightpath 2050" calls for -65% noise footprint, -75% CO_2 and -90% NO_x. Compared to the ambitions of the last century such as

- faster
- further
- higher

which had driven aerodynamic, propulsion, systems and structure technology to new horizons, climaxing in developments such as the Concorde, todays goals are quite different:

- more affordable
- more comfortable
- cleaner
- quieter
- safer

Safety has played a key role for commercial air transport ever since its very first days and is clearly also a psychological question of customer acceptance. However, it has to be stressed that today, compared to other transportation vehicles, aircraft already offers incomparably high safety standards, and the relevant certification rules have continuously been adapted. According to [7], 2110 persons were killed in car accidents, 9 persons in busses, and 3 persons in railways in the year 2009 in Germany. There were 0 fatalities in aircraft (MTOW > 5.7 t). In average for the years 2005–2009, 2.93 persons were killed statistically during car accidents every 10^9 km, 0.17 persons in busses, 0.04 persons in railways, 0.16 persons in trams and 0.00 persons in aircraft (MTOW > 5.7 t), Fig. 1.10.

Only two fatalities were recorded for scheduled service aircraft with more than ten seats in the U.S. during 2010 [8]. According to the National Transportation Safety Board NTSB, U.S. air carriers operating as scheduled service airlines have more than 17 million flight hours with more than 7 billion miles flown in approx. 9500 departures with no fatalities at all during the year 2010 [8]. On average, taking into account the years from 1992 to 2010, the statistical probability for a fatal

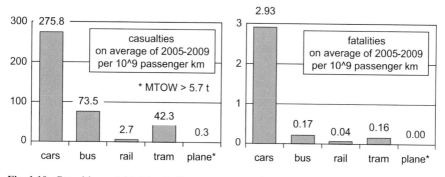

Fig. 1.10 Casualties and fatalities in German transportation systems [7]

accident per 100,000 departures is 0.017; this is equivalent to 1 killed passenger for approx. 6 Mio departures.

Even though these statistics impressively demonstrate the extremely high level of safety that has already been reached by modern aircraft design and air traffic control, engineers have to accept the demand for even higher safety. However, this means an additional challenge, as safety requirements can often be conflicting with economic and ecological requirements. To give only one example: For airframe structure design an increase of , allowing for higher sinking velocities in case of crash landings while protecting passengers from injury during touchdown and enabling fast and safe evacuation can easily mean a higher wall thickness of the affected structural parts, leading to more weight (and hence to higher fuel consumption and higher operating cost).

The oil price development since 1980 is shown in Fig. 1.11. Over the last 10 years the oil price has quintupled, and—even though the oil price is down to approx. US$30 at the time when this book is written—it is expected to keep rising again in the future. However, according to [1], while the traffic has increased by more than +50 %, the total jet fuel demand has almost been constant since the year 2000. This underlines the achieved efficiency increases of engines, aerodynamics, systems and light weight structures.

Highlights are fuel improvement and CO_2 efficiency achieved since the early jet planes in the 1950s [2] (Fig. 1.12). The enormous progress that has been achieved by aerodynamics, systems and structure engineers in designing advanced aircraft until today can also be visualised by defining the sum of airframe structure and system mass (without engines) needed for the transportation of one passenger over a range of 1000 km, Fig. 1.13.

It is evident that the ratio of mass and passenger-kilometres has been drastically improved in only a few decades and seems to be asymptotic since the 1980s. However, it must be stressed that at the same time the functionalities of commercial aircraft have been dramatically increased: where the first transportation aircraft usually only had a radio for the pilot, in-flight entertaining and communication

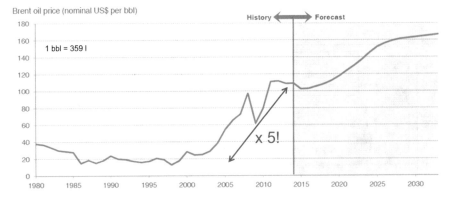

Fig. 1.11 Oil price development [1]

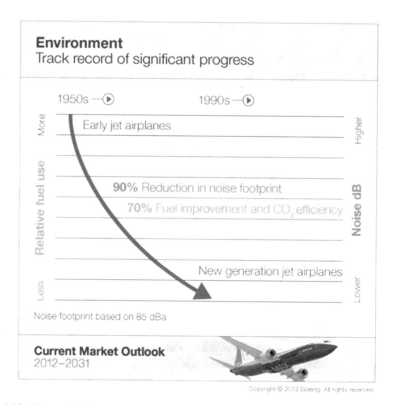

Fig. 1.12 Noise and fuel reduction [2]

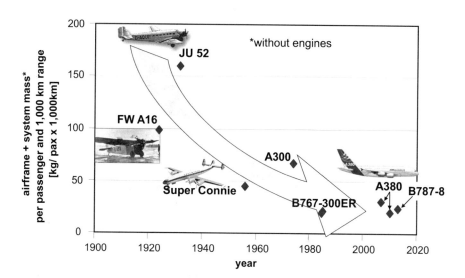

Fig. 1.13 Aircraft efficiency increase

systems for passengers today are considered as state of the art, to give only one example. As every additional function had to be fulfilled by some sort of (heavy) matter, more light weight design was needed in order to keep the value of about 25 kg of airframe and system mass per passenger and 1000 km range constant or to even undercut it.

Lighter Airframes: Contributors and Lever Arms

Main contributors for light weight airframe structures were

- improved understanding and predictions of loads
- active load alleviation systems
- advanced design principles
- improved design and sizing methods
- improved material properties
- improved manufacturing methods
- improved maintenance procedures

It is important to note that it is never a material with its properties alone, but it is always the smart combination of design, manufacturing and operational conditions that has to be considered in order to optimise weight savings. In addition to necessary improvements of systems, propulsion and aerodynamics, structural weight savings will continue to play a major role in the attempt to achieve even more overall efficiency and more ecological solutions for the next generation of aircraft. This is due to the high portion of mass attributed to the load carrying structure, Fig. 1.14. For an A320 aircraft, the load carrying structure makes more than 50 % of the weight (MWE, Maximum Weight Empty), followed by power plant, systems and furnishings.

Airframe structure is also very important regarding manufacturing cost. Disregarding engines, the airframe has the biggest portion of the manufacturing

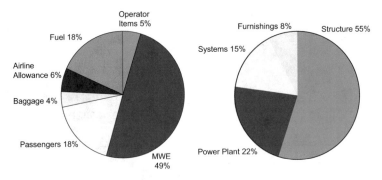

Fig. 1.14 Typical weight breakdown of a single aisle aircraft. *Left*: MTOW (maximum take-off weight). *Right*: MWE (maximum weight empty). Figures calculated from data given in [3]

cost. It is therefore key to develop material technologies which do not only allow weight savings, but also diminish manufacturing cost as much as possible.

Regarding the available "lever arms" of engineers designing a new aircraft, the definition of the right level of requirements is most important. On one hand, requirements must be set smart enough to guarantee sufficient attractiveness of the new aircraft for the market. On the other hand, hurdles must not be set too high, as there is a price to pay in terms of weight and manufacturer effort for each design solution fulfilling the requirement.

Some examples of very important requirements are shown in Fig. 1.15. A longer design service goal (DSG) or longer inspection intervals can mean additional material thickness (hence more weight, more cost). Higher passenger comfort by higher cabin pressure can lead to additional fuselage weight, coping with higher loads resulting from the higher pressure. The reduction of cabin noise can require thicker (and heavier) insulation or structural parts. Trades are necessary in the early phase of new aircraft developments in order to smartly fix the most important product requirements.

The next important "lever arm" is the configuration (i.e. shoulder wing aircraft, low wing aircraft or flying wing, cruciform tail or T-tail, wing mounted landing gear or body landing gear, rear or wing mounted engines, fuselage cross section, number and size of doors and windows, . . .). It can be a huge and costly effort for structure engineers to compensate the additional weight caused by the loads of rear mounted engines, or, to give another example, to compensate the additional weight of a non-circular fuselage cross section. New materials, adopted manufacturing technologies and material-optimised design solutions and function integration are also a very important lever arm for manufacturing cost and product performance (minimum weight, minimum operating cost).

There are numerous advantages of fibre reinforced plastics over conventional aluminium alloys (but also some important disadvantages, and these will be

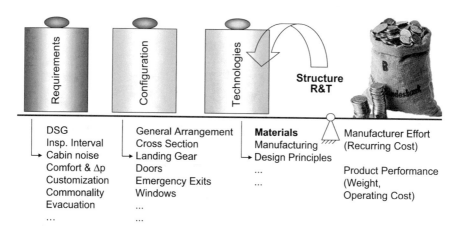

Fig. 1.15 "Lever arms" for efficient airframe design

discussed in the following chapters), that have contributed to the high portion of composites in modern airframe design. The most important are [9]:

- high structural performance to mass ratio
- utilisation of anisotropy for tailored design (in terms of strength, stiffness, stability)
- excellent media and corrosion resistance
- excellent fatigue behaviour ("no crack growth design concept")
- excellent fire resistance properties
- excellent energy absorption under compression

History of Fibre Reinforced Airframe Structure

Besides the fact that wood can also be considered an anisotropic fibre reinforced material and has been widely used for early airframe manufacturing since Karl Wilhelm Otto Lilienthal (23 May 1848 to 10 August 1896), the invention of phenolic resin by Leo Hendrik Baekeland (14 November 1863 to 23 February 1944) has to be seen as a key enabler for the introduction of fibre reinforced polymer composites in aircraft structures. The U.S. patent was filed in 1907, and the trade name for the product of the poly-condensation of phenol and formaldehyde was "Bakelit", manufactured by the Bakelite GmbH, later Bakelite AG from 1910 to 2004 (today Momentive).

Robert Kemp filed the patent "Structural Element" (U.S. Patent 1435244 A) in 1916 for the production of several airframe parts (including wings and fuselage) with fibre reinforced resins such as phenolic resin. There are many examples for successful aircraft with Bakelite resin applications. The Fokker F18 (1935) was a 12 passenger aircraft operated by KLM and had wings surfaced with Bakelite glued three ply. Horten manufactured a flying wing experimental aircraft (first flight 1937) with phenolic resin reinforced by embedded laminated paper, Fig. 1.16.

Fig. 1.16 Horten experimental aircraft with "Troilitax" laminates, consisting of paper reinforced phenolic resin [11]

Flax fibre reinforced phenolic resin was also used for airframe structures: "Gorden Aerolite" was used for the production of a fuselage of the Vickers Supermarine Spitfire in August 1940 [10].

Another important milestone was the invention and industrial manufacturing of epoxy resin. Pierre Castan applied for the patent of epoxy resin manufacturing in 1938 in Switzerland, which was filed in 1940. Paul Schlack in Germany applied for another patent in 1934, which was filed in 1939. Glass fibres were produced by Owens-Corning in the U.S. since the mid-1930s.

Other important patents were filed in 1940. British patent No. 505,811 and U.S. Patent No. 2,128711 on "Compound sheet material for the construction of aircraft and water vehicles" already claims reinforced thermoplastic matrix materials, although it took some time until high performance thermoplastics (such as PPS, already invented 1888 by Charles Friedel) could be produced on a larger industrial scale and successfully be applied for airframe applications, today known as "organo sheet material".

Another milestone was the application of Glass Fibre Reinforced Plastic GFRP for load carrying airframe structures. Wright Patterson designed a trainer aircraft "Vultee BT 15") with GFRP wing covers, Fig. 1.17; the span was approx. 13 m, and about 11,500 A/C were produced in total. The glass fibres were produced by Owens Corning.

The first "full composite airframe" was a sailplane developed by Akaflieg Stuttgart in Germany in 1954 (first flight 21 November 1957). Glass fibre reinforced polyester resin was used for wing, fuselage and tail of "fs 24 Phoenix", Fig. 1.18. The c_L/c_D ratio was 37.

As soon as carbon fibres could be produced on a larger industrial scale (first larger scale commercial production 1971 at Toray in Japan), these materials could also be used for airframe design. The first application for a large civil transportation aircraft was the fin (vertical tail plane) of Airbus A310-300 in 1985, which is often referred to in literature. However, other important parts, relevant for aircraft control and flight safety, had already been introduced by Airbus in 1982 for the A310-200, Fig. 1.19. Spoiler, rudder and elevators were composite structures. For A320 (1987), not only spoiler, rudder, elevator and fin were composite, but also main

Fig. 1.17 Wright Patterson trainer aircraft "Vultee BT 15" with GFRP wings [12]

Fig. 1.18 GFRP glider "fs 24 Phoenix" [13]

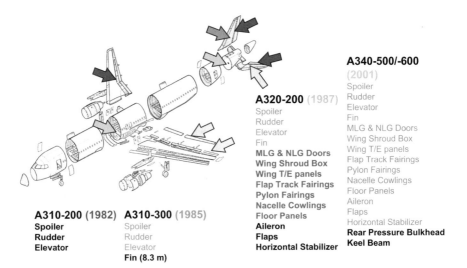

A340-500/-600
(2001)
Spoiler
Rudder
Elevator
Fin
MLG & NLG Doors
Wing Shroud Box
Wing T/E panels
Flap Track Fairings
Pylon Fairings
Nacelle Cowlings
Floor Panels
Aileron
Flaps
Horizontal Stabilizer
Rear Pressure Bulkhead
Keel Beam

A320-200 (1987)
Spoiler
Rudder
Elevator
Fin
MLG & NLG Doors
Wing Shroud Box
Wing T/E panels
Flap Track Fairings
Pylon Fairings
Nacelle Cowlings
Floor Panels
Aileron
Flaps
Horizontal Stabilizer

A310-200 (1982)   **A310-300** (1985)
Spoiler   Spoiler
Rudder   Rudder
Elevator   Elevator
                       Fin (8.3 m)

Fig. 1.19 History of CFRP introduction to Airbus aircraft [14], detailed information can be collected from [3]

landing gear (MLG) and nose landing gear (NLG) doors, the wing shroud box, wing trailing edge panels, flap track fairings, pylon fairings, nacelle cowling and floor panels consisted of carbon fibre reinforced plastics CFRP.

In addition, ailerons, flaps and the horizontal stabiliser were also designed and manufactured in CFRP. Aramid fibre reinforced plastics AFRP was used for the radome, and glass fibre reinforced plastic was used for some parts of the belly fairing and the leading edge of the vertical tail plane. In 2001, first fuselage applications were introduced for A340: The rear pressure bulkhead, closing the pressurised passenger compartment towards the rear, was designed and manufactured in CFRP. A CFRP design was also chosen for the keel beam.

Airbus continued the "step by step" principle of CFRP introduction to primary load carrying airframe structures with the A380 development. For A380 (first flight 2005), the complete centre wing box, connecting left and right wing to the forward and rear fuselage, was a CFRP design. In addition, CFRP was also chosen for the

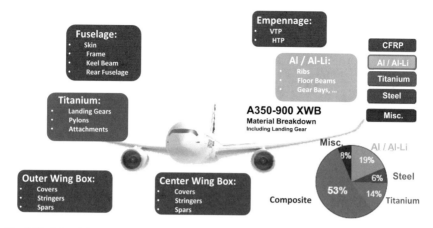

Fig. 1.20 Material mix of Airbus A350XWB [15]

rear fuselage behind the rear pressure bulkhead, the passenger floor beams in the upper deck, wing ribs and the wing J-nose.

The first Airbus full CFRP wing was designed for A400M (first flight 2009), with a span of approx. 42 m. Although Boeing was also using CFRP for primary load carrying airframe structures in commercial aircraft, Airbus was considered to lead the technology for a long time. With the B787 "Dreamliner" (first flight 2009), however, Boeing was the first manufacturer to introduce a "full composite" airframe, since not only tailplane and wing structure, but also the complete fuselage structure came as a CFRP design. Airbus followed this development with its A350XWB. The material mix is shown in Fig. 1.20.

Carbon Fibre Consumption

The content of CFRP in airframe structure is more than 50 % for the latest Boeing and Airbus aircraft, Fig. 1.21.

This has also contributed to the increasing demand of carbon fibres and the global demand of carbon fibre composites. According to [16], the global demand of carbon fibres has increased from 26,500 t in the year 2009 to around 60,000 t in the year 2015 (for comparison: aluminium approx. 53 million tons, steel approx. 1.7 billion tons). At the same time, the CFRP consumption doubled, Fig. 1.22.

Since a higher demand is also expected for automotive carbon fibre applications, some manufacturers of carbon fibres are currently increasing their capacities. At present, the global carbon fibre manufacturing capacity is estimated at over 100,000 t, [16]. Comparing different market sectors and revenues, aerospace is clearly dominating, Fig. 1.23. This must also be attributed to military CFRP applications.

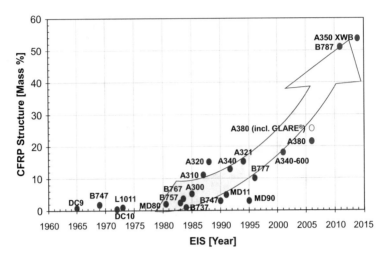

Fig. 1.21 Content of CFRP in airframes

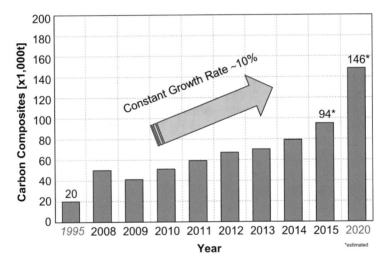

Fig. 1.22 Global CFRP demand [16]

End of Life

As more and more aircraft are reaching their design service goal (often after 20 or 30 years of service), recycling becomes more and more important. In a first step, all liquids have to be disposed (water, fuel, hydraulic fluid, lubricants etc.). The second step comprises the removal of avionic, landing gears, engines and auxiliary power units. In a third step, all cabin items and linings, insulation materials etc. must be removed. In the last step, the airframe structure can systematically be dismantled, and different materials (steel, aluminium, copper, titanium, CFRP, etc.) can be separated

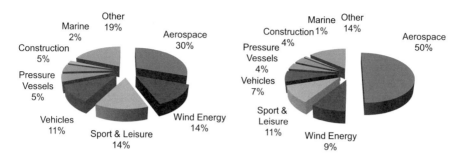

Fig. 1.23 CFRP market shares by mass (*left*, 100 % = 46.5 kt) and revenues (*right*, 100 % = US $1.7 billion), Figures for the year 2013 collected from [16]

Table 1.1 Energy consumption and carbon footprint of different materials; energy and CO_2 values from [20], all other values are own estimations

Resource	Unit	Carbon fibre	Epoxy resin	Aluminium sheet	Steel sheet
Energy consumption for production	[MJ/kg]	200...400	130	150	20
CO_2-equivalent	[kg CO_2/kg material]	10...20	7	12	3
Production cost	[euros/kg]	20...100	50	15	1

from each other for a subsequent closed loop recycling management. The material recycling of CFRP is still in its very beginnings. General procedures are described in [17]. A good overview of the state of the art at Airbus can be found in [18].

The dominating procedure for cured parts or end of life CFRP parts is crushing and milling, with a reuse of relatively short fibre fragments in lower grade applications, [19].

Since the production of carbon fibres is very energy intensive, Table 1.1, large efforts are presently undertaken to recover the fibres from old CFRP structures in full length, avoiding any "down cycling", i.e. a reduction of material properties for the new application. However, technologies allowing to economically separate fibres from the cured epoxy resin, spinning high quality staple fibre yarn out of recycled carbon fibres, and technologies for semi-finished material production such as unidirectional recycled carbon fibre prepreg are not yet fully developed, and still a matter of research.

Questions

1. What are the main demands on the transport system aircraft?
2. Why does mass play such an important role for the fuel consumption?
3. How can the balance of forces at the centre of gravity of an aircraft be described in a simplified manner?

4. Which fuel savings could be achieved by reducing the structural mass of a commercial single aisle aircraft by 1 kg?
5. How many commercial aircraft are currently being produced each year?
6. How many commercial aircraft are currently in use worldwide?
7. How many passengers participate each year in the world's aviation?
8. What has the growth rate been for worldwide passenger traffic in recent decades?
9. Into which segments can the commercial aircraft market be divided?
10. Which demand for commercial aircraft is expected in the next 20 years? What are the key factors?
11. In which aircraft market segment can we expect the largest quantities?
12. What were the goals of aircraft design in the past? What are the main challenges of the European Aeronautics Vision 2020?
13. What are typically the larger shares of operating costs of a commercial aircraft, which have a major impact on the choice of structural materials in particular?
14. Importance of lightweight construction: What is approx. the share of fuel costs in the total operating cost of a single aisle commercial aircraft?
15. What important shares of total mass of the aircraft exist? Which part has the largest share of the take-off weight of an aircraft?
16. How could the structural mass of commercial aircraft be gradually reduced over time?
17. Which important factors influence the production costs and operating costs of a commercial aircraft? Give examples.
18. What value (in € or US$) has a saved kg of structural mass for commercial aircraft? Would you expect a different value for automobiles? If yes, what causes the difference?
19. Aircraft lightweight components: What are key benefits of CFRP compared to aluminium?
20. More and more components were successively made of composite materials. What were the most important steps, and in which order were they taken?
21. What percentage of the structural mass is made from composite materials in modern-day commercial aircraft?
22. How many tonnes of CFRP are produced p.a.?
23. Name a few carbon fibre manufacturer.
24. Why is it so important to recycle carbon fibres at the end of the product life cycle?

References

1. Flying on Demand, Airbus Global Market Forecast 2014–2033, AIRBUS S.A.S. 31707 Blagnac Cedex, France, 2014. http://www.airbus.com/presscentre/corporate-information/key-documents/
2. Boeing Current Market Outlook 2015–2034. http://www.boeing.com/commercial/market/

3. Jackson, P., Bushell, S., Willis, D., Munson, K., Peacock, L.: Jane's All the World's Aircraft 2011–2012. Jane's information group, ISBN-10: 0710629559
4. US DOT Form 41 Airline Operational Cost Analysis Report. IATA (International Air Transport Association), Airline Operational Cost Task Force (AOCTF) (March 2011). http://www.iata.org/whatwedo/workgroups/Documents/aoctf-FY0809-form41-report.pdf>
5. European Aeronautics: A Vision for 2020. Meeting society's needs and winning global leadership. Report of the Group of Personalities, January 2001. European Commission, Luxembourg, Office for Official Publications of the European Communities (2001). ISBN 92-894-0559
6. Flightpath 2050, Europe's Vision for Aviation. Report of the High Level Group on Aviation Research. European Commission, Luxembourg, Publications Office of the European Union (2011). ISBN 978-92-79-19724-6, doi:10.2777/50266
7. Statistisches Bundesamt, Wirtschaft und Statistik 12/2010, Grafik 2010-01-0868 bzw. 2010-01-0869) Unfallstatistik – Verkehrsmittel im Risikovergleich, Statistisches Bundesamt, Wiesbaden
8. National Transportation Safety Board NTSB. Review of US Civil Aviation Accidents (2010). http://www.ntsb.gov/data/aviation_stats.html
9. Neitzel, M., Breuer, U.: Die Verarbeitungstechnik der Faser-Kunststoff-Verbunde. Carl Hanser Verlag, München (1997)
10. Unckenbold: Lightweight Materials with Long Fibre Reinforcement – Innovative for 80 Years. International AVK Conference 2011
11. Horten, R., Selinger, P.: Nurflügler – die Geschichte der Horten- Flugzeuge 1933 bis 1960. H. Weishaupt Verlag, Graz. ISBN: 978-3-900310-09-7, 7. Aufl.
12. Flight, 10 April 1947, page 313
13. www.uni-stuttgart.de/akaflieg
14. Breuer, U.: Herausforderungen an die CFK Forschung aus Sicht der Verkehrsflugzeug Entwicklung und Fertigung, 10. Nationales Symposium SAMPE Deutschland e.V, Darmstadt (2005)
15. Hellard, G.: Composites in Airbus. EADS Global Investor Forum (January 2008)
16. AVK & CCeV Composites Market Report (2014)
17. Neitzel, M., Mitschang, P., Breuer, U.: Handbuch Verbundwerkstoffe, 2. Aktualisierte und erweiterte Auflage. Carl Hanser Verlag, München (2014)
18. Witte, T., Munoz, P.: CFRP Reuse and Recycling Strategies for Material-Efficient Composite Production. 1st International Composites Congress, 22. September 2015, Stuttgart
19. Kreibe, S.: Recycling of CFRP – A Promising Challenge for a Delicate Material. 1st International Composites Congress, 22. September 2015, Stuttgart
20. Leichtbau in Mobilität und Fertigung, Ökologische Aspekte. e-mobil BW GmbH, Stuttgart (2012). www.e-mobilbw.de

Chapter 2
Requirements, Development and Certification Process

Abstract The airframe material selection is decisive to determine target weights as well as manufacturing and operating cost of aircraft components. It is a prerequisite for production planning and long lead time items. In addition, it is key for identifying any material-related certification risks and for establishing the certification process planning and means of compliance. It is therefore very important to freeze the material decision for components well before a new aircraft is offered to the market. This chapter starts with a discussion of the development process of new aircraft and the definition of the basic milestones. Fundamental aircraft requirements of operators, authorities and airframe manufacturers influencing the material decision are highlighted, and the requirement cascade is introduced. The functional analysis method as part of the design process is explained followed by the description of the general procedure of the structure stressing and certification process. Special emphasis is put on the description of the "no crack growth" concept of CFRP structures.

Keywords Development of new aircraft • Top level aircraft requirements • Aircraft design goals • Requirement cascade • Design process VDI 2221 • Functional analysis • Design service goal • Limit load • Ultimate load • Safety factor • Reserve factor • Design load cases • Structure stressing • Certification process • Damage tolerance • Crack growth • No crack growth concept • Safe life • Fail safe • Impact • Compression after impact

Development Process and Requirement Cascade

Although graphs similar to the one shown in Fig. 2.1 have been and still are discussed in many textbooks, it cannot be stressed enough how important it is to use the early phase of a new aircraft development in order to carefully examine and smartly define product requirements. As the freedom of choice is still relatively high in the concept phase, large efforts have to be undertaken in order to trade different concepts and to find the best compromise to cope with operator, regulator (authorities) and manufacturer requirements. Towards the end of the concept phase more than 2/3 of the manufacturing costs are already fixed, and the freedom of choice rapidly decreases during the definition phase, when manufacturing drawings

© Springer International Publishing Switzerland 2016
U.P. Breuer, *Commercial Aircraft Composite Technology*,
DOI 10.1007/978-3-319-31918-6_2

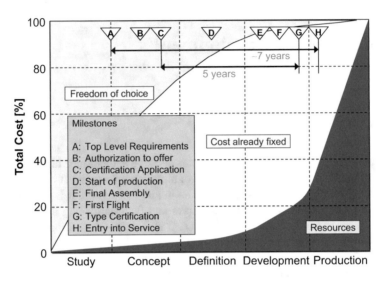

Fig. 2.1 Development phases and cost allocation

are being generated for all parts. In practice, this is also the time in which deltas are detected by detailed weight analysis based on CAD-models and supporting FEM-calculations, often indicating a weight above the target. Changes from the originally selected material to a lighter one (in order to reduce weight) are very difficult in such a late stage of the development process. Material qualification is a very time consuming and expensive process, and material changes can also impact production technology, for which long lead time items (tooling, manufacturing facilities etc.) are needed.

Providing a high amount of engineering resources as well as expensive test material and test equipment is necessary when developing a new civil transportation aircraft for commercial operation. It also requires the provision of a high amount of capital: The total development cost of a new transportation aircraft can add up to 10 billion euros. In order to limit the financial risk linked to this investment, aircraft manufacturers have a high interest in selling their new aircraft long before the first plane is tested and certified. In fact, the "authorisation to offer", signing contract orders with airline operators, usually takes place during the concept phase, some 7 years before entry into service, Fig. 2.1. The customer (the airline) buys a product that only exists on paper, and the seller (the aircraft manufacturer) guarantees a certain product performance without having defined, built or tested real design solutions for the aircraft. Hence, the risk of failing to meet important product requirements, probably resulting in price deductions, penalties or even order cancellation, must and can only be reduced by the intensity and quality of early concept studies and trades. This requires highly skilled experts.

The three main parties defining the most important requirements are the regulator (i.e. the government authorities, in Europe the European Aviation Safety Agency EASA, and in the U.S. the Federal Aviation Administration FAA), the operator (airlines) and the aircraft manufacturer, Fig. 2.2. The type certification of a new aircraft is provided by the regulator, and aircraft manufacturers are obliged to demonstrate that their new aircraft complies with all applicable rules. Safety standards are set by the authorities, as well as environmental standards. The manufacturer's goal is providing an attractive and successful product to the market; minimising manufacturing cost is a key requirement. In consequence, manufacturers are interested in using existing infrastructure and existing resources as much as possible. The operator is seeking product attractiveness for passengers (or freight) at minimum total cost of ownership; a low purchase price, low fuel and maintenance cost, high passenger comfort, maximum payload, range flexibility, cabin flexibility, superior take-off, landing and manoeuvring performance, a long service life (high resale value) and many more advantages are desired in particular. These requirements of regulators, operators and manufacturers can easily conflict with each other. The lightest material, enabling very low airframe mass and low fuel consumption, will very likely also be the most expensive material, leading to high manufacturing cost, to give only one example. High safety standards—a must for every airframe design—can also be a matter of high operating cost when affecting maintenance or repair. After all, a new aircraft, having to fulfil regulatory requirements for its certification, will always be the best compromise between operator and manufacturer requirements. This also means that requirements—most likely increasing downwards along the requirement cascade—are prone to changes and adaptions throughout the complete development process.

In order to capture, document and elaborate requirements, guide the design, validate design solutions and verify hardware (i.e. performing tests, demonstrating that requirements are fulfilled), all requirements are carefully recorded and organised in a cascade, Fig. 2.3.

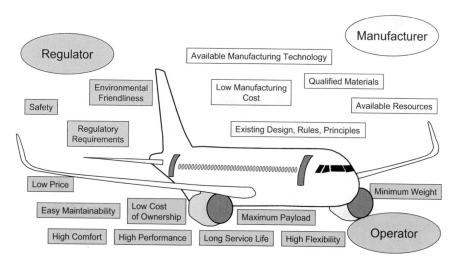

Fig. 2.2 Typical design goals of regulator (authorities), manufacturer and operator

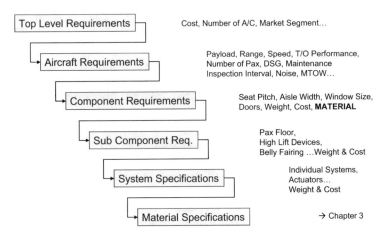

Fig. 2.3 Requirement cascade

The requirement cascade usually starts with top level requirements. These include the market segment (i.e. short range, medium range or long range aircraft), the number of aircraft to be produced, the cost etc. . . The aircraft requirements (see milestone A in Fig. 2.1, at which these requirements should be frozen) define for example the payload (number of passengers, weight or freight), range, speed, take-off and landing performance, design service goal (number of flights, total flight hours, years of service), maintenance, inspection intervals, noise etc. . . Any aircraft can be divided into its large components: wing, fuselage, tail plane, landing gear, engines. Examples for typical component requirements for a fuselage are diameter, length, seat pitch, aisle width, window size, doors etc. . . Weight and cost targets are defined on aircraft as well as on component and sub component level, down to individual parts. Material requirements can already be defined on component level, too, as the choice of material has a high impact on total aircraft cost as well as on development and production effort. Components can be divided into sub components. Typical sub components for the fuselage are the passenger floor, the cargo floor, the belly fairing etc. . . The requirement cascade continues with specifications for individual parts or systems. Finally, requirements for materials are a matter of the material qualification process, as discussed in Chap. 3.

The compilation, definition and documentation of requirements are extremely important. Many delays, unexpected high development costs, problems of industrial production ramp up, failures to meet contracted performance guarantees etc. are directly linked to missing, incorrect or conflicting requirements. Other complications may include non-functional yet solution-oriented requirement definitions, inadequate monitoring processes of design stages by validating intermediate design solutions, carefully analysing the fulfilment of requirements by the proposed solution, etc. . .

Design Process According to VDI 2221

An appropriate method for a structured design process is defined by VDI guideline 2221 [1] (VDI = Verein Deutscher Ingenieure), Fig. 2.4.

One of the most important steps is the precise definition of functions. In order to guarantee a thorough understanding of the product requirements at the very beginning of its development, functional analysis must be organised involving experts from all disciplines (structure, systems, aerodynamics). Functional analysis enable to think in terms of functions and not in terms of solutions. The precise definition of functions also helps understanding and identifying the root cause of unwanted intermediate product features (i.e. in case it is discovered that a design solution for a given part would lead to too much weight or cost) and consequently adapting the design solution, or challenging (and potentially changing) a requirement in order to meet more important requirements from superior levels of the requirement cascade. A good example for a purely functional requirement is: "The floor must be able to carry payload (maximum load 32 t)". A bad example (as it is already solution oriented) in this context would be: "The floor structure is an aluminium honeycomb design capable to withstand a maximum load of 32 t".

The basic procedure for a functional analysis is described in Fig. 2.5.

After a description of the product (step 1), the functions are systematically analysed (step 2) by a team of experts from different disciplines. Step 3 is the hierarchical organisation of the different functions, while step 4 characterises all functions. The final step prioritises these functions. An example is illustrated in

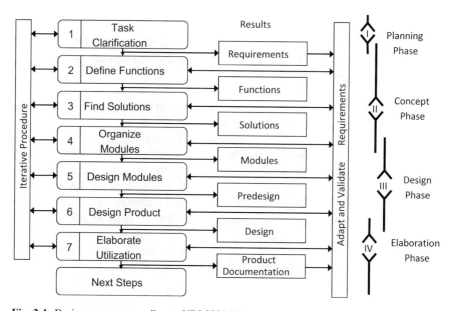

Fig. 2.4 Design process according to VDI 2221 [1]

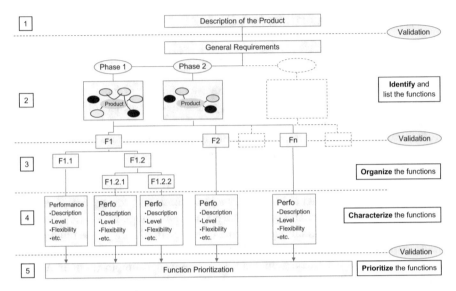

Fig. 2.5 Functional analysis process

Fig. 2.6. The product to be designed is depicted in the centre (step 1), surrounded by all elements that could interfere with the product throughout its complete life span. This requires the analysis to be individualised for all phases: assembly, handling, transport, test, operation, maintenance and disassembly + recycling.

In step 2, any important interaction between the surrounding elements and the product has to be analysed and recorded in a purely functional manner, i.e. as free of existing or new design solutions as possible. Some possible examples are given in Fig. 2.7. Once the functions are identified, they are organised within functional tree architecture (step 3) in order to facilitate the reading and understanding, and to identify sub or lower level functions, Fig. 2.8.

Step 4 deals with the characterisation of each individual sub (or lower level) function. Characterisation means that quantifiable criteria have to be defined, enabling designers to implement these characteristics in models and drawings, and enabling engineers to validate that requirements have been properly considered. Tables as shown in Table 2.1 can be used to record these criteria as well as to provide references to documents. They can also be used to record the result of step 5 of the functional analysis, the prioritisation. For this purpose, flexibility levels can be assigned to each criterion: F_0 could be used to indicate that the criterion is a must, F_1 could indicate flexibility, F_2 could indicate that the relevant criterion is just something "nice to have". In practice, if at some stage of the design it turns out that the predefined solution is violating individual requirements, it is often a chief engineer's decision whether a redesign is necessary to fully meet the requirement or if criteria for requirements are adapted.

In order to transfer functional requirements into design drawings, following the principal procedure described in Fig. 2.4, different design solutions for the

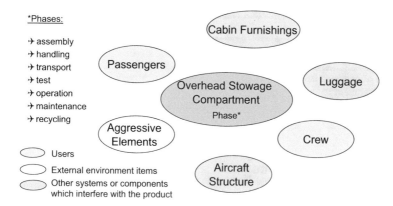

Fig. 2.6 Functional analysis example

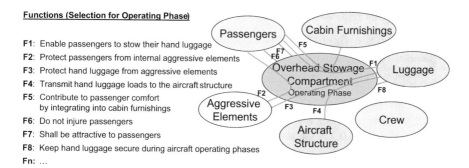

Functions (Selection for Operating Phase)

F1: Enable passengers to stow their hand luggage

F2: Protect passengers from internal aggressive elements

F3: Protect hand luggage from aggressive elements

F4: Transmit hand luggage loads to the aircraft structure

F5: Contribute to passenger comfort by integrating into cabin furnishings

F6: Do not injure passengers

F7: Shall be attractive to passengers

F8: Keep hand luggage secure during aircraft operating phases

Fn: ...

Fig. 2.7 Selection of possible functional requirements for the operating phase

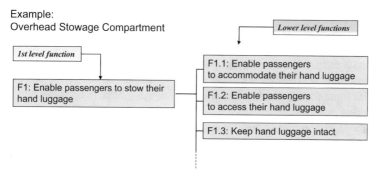

Fig. 2.8 Functional tree

functions can be traded, selected and combined by means of morphological boxes. The process is described in [2].

Table 2.1 Typical results of function characterisation (selection of examples, figures not to be taken for design)

Criteria	Level	Reference	Flexibility	Means of control
Volume of overhead stowage compartment	170 dm^3 ± 5 % or 56 dm^3 per passenger	Master geometry	F2	Design review
Shape of overhead stowage compartment	Parallelpiped constant thickness max. 25 cm	Master geometry	F0	Design review

Load Cases and Stressing

An important part of defining and refining design schemes is stressing. External loads are needed for calculating internal structural stresses and for sizing thicknesses (or to re-arrange the topology) for minimum weight accordingly. It is necessary at this stage to define some important terms:

- *Design Service Goal* (DSG) is an important aircraft requirement. It is the planned life time of the aircraft. The figure is provided in flight cycles (FC), years or flight hours (FH), and relevant for design an certification
- *Limit Load* (LL) is the maximum load to be expected by the airframe throughout the complete service life (DSG)
- *Safety Factor* (SF) is a factor applied for airframe design (usually 1.5)
- *Ultimate Load* (UL) is the product of LL and SF
- *Reserve Factor* (RF) is the relationship of structure strength (sustained load) and ultimate load. It should be ≥1.0.

The design service goal depends on the type of aircraft and the main missions it is designed for. The A320 was originally designed for a design service goal of 20 years or 48,000 flight cycles. The A380 was designed for only 19,000 flight cycles; however, the typical duration of a flight is much longer compared to the A320. The design service goal is not a life limit: Life extension is possible by means of adequate fatigue test results and positive in-service experience. Special maintenance programs are defined in case of a life extension. The design service goal is an important input for the assumption of load cycles that have to be taken into account for stressing purposes.

Steady load conditions and incremental loads of different phases have to be analysed. The most important phases are "on ground" and "in flight", Fig. 2.9, but there are other phases which can also be very important for structure design: During manufacturing or assembly, tool drops on structural parts can cause large impact energies and have to be taken into account for design and stressing. Ground load cases can be linked to braking, turning, freight loading and unloading operations etc. and have to take into account worst case conditions (i.e. temperature influences, media influences etc.). Typical flight load cases occur during manoeuvring and during gusts, but cabin pressurisation is also an important case. Accidental damages

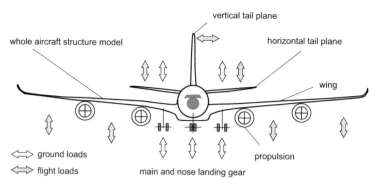

Fig. 2.9 Aircraft model and loads

and the resulting loads must also be assessed (for example depressurisation of the cabin or a tyre burst with broken parts impacting the fuselage).

To support aircraft manufacturers and to make sure that all important and safety relevant load cases are properly taken into account, the regulators (authorities) provide a catalogue of requirements applicable for certification. In Europe, this is the Certification Specification [3], in the United States it is the Federal Aviation Requirements (FAR 25) [4]. As the regulations defined by the authorities are also prone to updates and changes, the applicable set of requirements is fixed at the point in time when the aircraft manufacturer officially applies for the certification of a new type of aircraft being developed. This is usually linked to the milestone "authorisation to offer", Fig. 2.1, and the following board decision of the aircraft manufacturer to launch the programme or not, depending on the number of aircraft sold to the first customers. From this point in time, a maximum of a 5 years period is allowed to demonstrate that all requirements can be fulfilled for the type certification.

The general procedure for structure stressing and certification is:

(1) Definition of the aircraft geometry and its outer loft
(2) Generation of an aircraft finite element model
(3) External loading of the model with all relevant cases
    as defined by the rules of the authorities, including safety factor application
(4) Calculation of forces, displacements and internal stresses/strains
(5) Comparison of sustainable stresses (depending on material allowables under the given temperature and media influences, wall thicknesses, topology, etc.) to internal stresses. This is supported by tests. Manufacturing and in-service damages have to be taken into account.
(6) Calculation of reserve factors
(7) Re-design for minimum structural weight

As the shape and mass of the aircraft are prone to changes during the development and are both influencing the loads, load loops can be necessary and lead to a redesign of parts, Fig. 2.10.

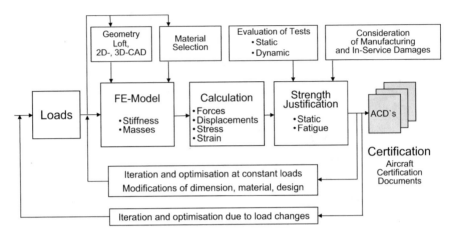

Fig. 2.10 General procedure of stressing and certification process

The goal of a thorough analysis is to calculate all relevant load cases for the part being designed, and to find out which load case is the most relevant one, i.e. leading to the maximum internal stresses and thus defining the dimensions of the part. Figure 2.11 illustrates some design load cases and resulting stresses.

Especially in the early phase of the development process these findings are very important for the selection of "the right material at the right place".

The basic structural certification approach is based on analysis validated by testing . Each critical loading condition is analysed in order to demonstrate compliance with strength and deformation requirements. A "test pyramid" is used to provide results at different levels of the design and development stage, and to verify that the structure fulfils all requirements, Fig. 2.12. The pyramid will be discussed in Chap. 5.

Damage Tolerance Requirements: The "No Crack Growth" Concept of CFRP

The applicable rules for relevant load cases, safety factors etc. defined by the authorities can only reflect the latest best available technology ("state of the art"). Since the new aircraft can contain new technology such as forward looking gust sensors and active load alleviation systems, structural health monitoring systems, new types of engines, unconventional wing architecture etc., it is necessary to assess any kind of certification risk already before the "authorisation to offer" and to discuss possible mitigation technologies with the authorities. It is also necessary to agree with the authorities on special analysis and tests to be carried to demonstrate that all risks have been properly assessed and can be mitigated by adequate design solutions or additional precautions.

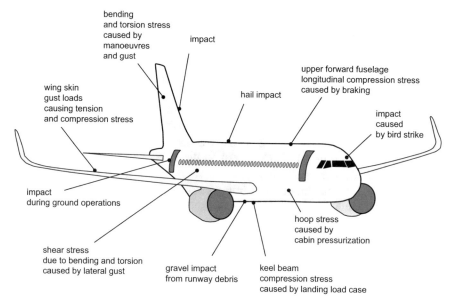

Fig. 2.11 Examples for design load cases and resulting stresses

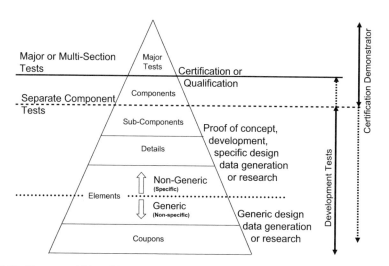

Fig. 2.12 Test pyramid (see Chap. 5)

Safety requirements have been revised and improved continuously [5]. Very important steps were

- Introduction of "safe life" requirements in the 1940s
- "Fail safe" requirements in the 1950s
- "Damage tolerance" requirements in the 1970s

"*Safe life*" means that the structure has a safe, limited lifetime. Analytical and experimental verifications with appropriate safety factors have to be demonstrated.

"*Fail safe*" means that different load paths are available. In case one load path fails, the remaining load path can sustain the load.

"*Damage tolerance*" means the capability of a damaged structure to sustain the loads until the damage is detected and repaired during a scheduled inspection (or if the damage leads to a non-critical loss of a function).

The damage tolerance requirements were revised following in-service incidents. A famous incident in 1954 was the explosive fuselage decompression of the De Havilland Comet 1 (DH 106-1, G-ALYP) on the 10th of January after only 1286 pressurised flights and less than 2 years of service, and the disintegration of a second Comet 1 on the 8th of April in the same year. An intensive investigation revealed metal fatigue as the main cause of these accidents [5]. As a consequence, the authorities demanded a certification either by fatigue evaluation of the airframe ("safe life" approach) or by a "fail safe" design, demonstrating that after failure of a single principal structural element catastrophic failure would not be probable.

Incidents in the 1970s (loss of an AVRO 748 in Argentina 1976, Dan Air B707 crash in Zambia 1977), the damage tolerance concept was included into the fatigue evaluation of the structure by the FAA [5].

Another very famous incident was Aloah Airlines flight 243 on April 23rd, 1988. A 19 year old Boeing 737 with 35,500 flight hours experienced a fuselage panel failure in 24,000 ft. The pilot managed to land with 94 survivors and only 1 fatality (a crew member). The result of the following examination by the National Transportation Safety Board revealed corrosion and multiple fatigue cracks, and the subsequent failure of riveted fuselage panel joints as the key origin of the incident (Fig. 2.13).

To prevent this kind of failure, a mandatory design process for damage tolerant structures was introduced:

- Aloah-Airlines Flight 243
- Boeing 737
- >90.000 F/C
- 28 April 1988
- fuselage panel failure in 24,000 ft
- pilot managed to land
- 1 fatality
- 94 survivors
- origin: multiple fatigue cracks, corrosion, failure of riveted joints

Fig. 2.13 Aloah Airlines Flight 243, image source see [6]

(1) Identification of structure critical elements
(2) Definition of the types of damages to be encountered
(3) Assessment of damage initiation time, propagation rate, critical size
(4) Definition of minimum detectable size
(5) Definition of inspection threshold, inspection interval, access and inspection method

Especially after the ALOAH accident, the authorities insisted on including manufacturing defects as a damage source. This has led to more safety, but also to additional development efforts as well as additional maintenance efforts (and costs!) in particular for conventional aluminium airframe structures.

The growth of a crack within an aluminium sheet is schematically depicted in Fig. 2.14. At a certain time (after a certain number of flight cycles), the crack reaches a length which reduces the residual strength of the structure to limit load. It has to be ensured by appropriate design (i.e. the selection of the right material, the sizing to an adequate thickness), that a crack detected during an inspection will not reach its critical length before the next scheduled inspection, Fig. 2.15. The length of the inspection interval also depends on the minimum detectable crack length. Typical values for the minimum detectable crack length (Visual Special Detailed Inspection VSDI) are between 35 mm and 70 mm, depending on light conditions and surface. Using special test equipment (High Frequency Eddy Currents HFEC), applied by specially trained and qualified personnel only, it is possible to detect very small cracks down to 1 mm length.

For CFRP, the "no crack growth" concept can be applied. If the maximum strain of a dynamically loaded CFRP structure is limited to a certain value (depending on the type of material typically to 0.4 %), cracks (delamination) within the structure,

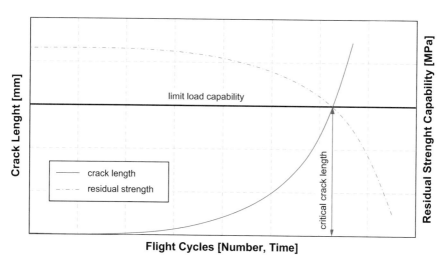

Fig. 2.14 Crack growth within aluminium sheet and residual strength capability

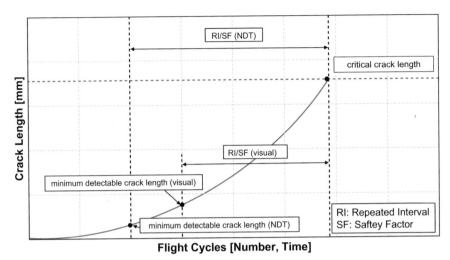

Fig. 2.15 Crack growth and inspection interval for an aluminium skin

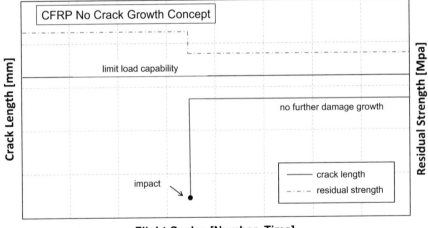

Fig. 2.16 Crack growth (usually delamination size growth) and residual strength capability of a CFRP sheet

as they can be caused by impact events, for example, will not grow. Figure 2.16 also refers to [7].

This CFRP design means a key advantage over classic aluminium designs, as it reduces the maintenance effort for the operating airline. Reduced maintenance time means less time in the hangar and more time to transport payload and create revenue.

As there are several impact events that can occur during aircraft service, Fig. 2.17, but even already during manufacturing and assembly of the aircraft, it

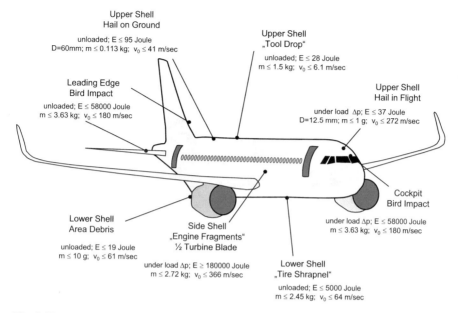

Fig. 2.17 Impact events (examples only, not to be taken for design, based on [8] and [9])

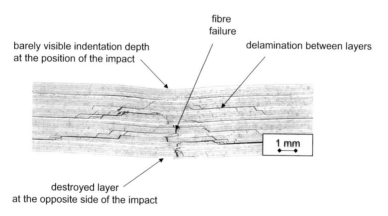

Fig. 2.18 Barely visible CFRP impact damage. The structure looks intact and faultless from above, although it contains severe damages inside

is necessary to take these impacts into account during the structure development, and to verify the "no crack growth concept" for CFRP by analysis and test.

Since impact events are also possible for CFRP, where damage occurs within the laminate structure (matrix failure, delamination, fibre failure) while no damage is visible from the outside, Fig. 2.18, it is necessary to design the structure in a way that provides ultimate load capability for these cases.

This damage tolerant CFRP-design, considering not only manufacturing imperfections but also impact damages below the visibility threshold, will allow a

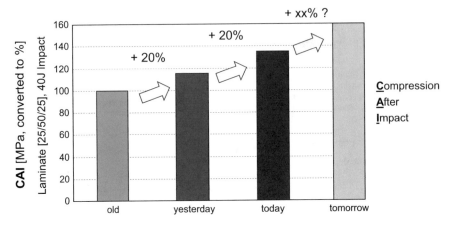

Fig. 2.19 Compression after impact strength, comparison of "old" 125 °C curing epoxy systems and new, toughened 180 °C curing prepreg materials

"maintenance friendly" aircraft structure in service and limit repair efforts of the operator: Impact damage that cannot be detected from the outside will not require any repair for the complete design service goal, see Chap. 6. On the other hand, this customer-friendly requirement of robustness can also come with a price in terms of weight, as a minimum wall thickness is necessary to cope with probable impact events. CFRP material manufacturers have constantly and successfully improved impact damage tolerance by toughening the epoxy resin, Fig. 2.19, and see also Chap. 3.

However, attempts to improve the impact damage tolerance by z-reinforcements, i.e. out-of-plane reinforcements such as pins, special yarns etc., have not been widely introduced into series applications so far, as these reinforcements degrade in-plane material properties with negative effects on the part weight.

Figure 2.20 shows the detectability (dent depth) of an impact versus the impact energy [10]. For realistic energy levels and below the detectability threshold, the structure must sustain ultimate load, following the concept of robustness. An impact event is assessed as realistic if it occurs once within 10^5 flight hours. The detectability threshold for a visual inspection of a CFRP airframe structure is typically about 0.1 mm dent depth. Up to extremely improbable impact events with high impact energies and an occurrence of one event within 10^9 flight hours (for comparison, the design service goal of A320 was 48,000 flight cycles; for typical 2 h flight missions this means less than 100,000 flight hours) and up to large visible impact damages, at least limit load must be sustained.

The size of "large" visible impact damages is defined by certification specification requirements, in case of new technologies and associated risks it has to be agreed with the authorities for the relevant structural parts. For a "thin" CFRP part, the detectability threshold is already reached at relatively low energy levels with a

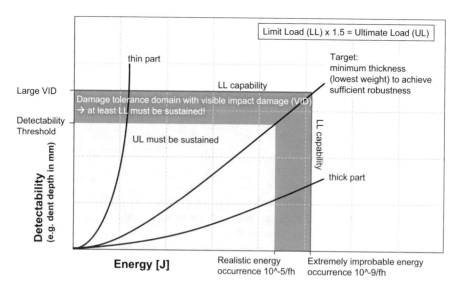

Fig. 2.20 Detectability of impact damages versus impact energy [10]

high probability of occurrence. This kind of light weight design would require frequent repair and is not favourable from an airline operator point of view. A "thick" part would never reach the detectability threshold (and thus never require any repair), but this over fulfilment of robustness would lead to a heavy design solutions. The optimum solution is the thickness that is just achieving enough robustness and sustains the loads for damages caused by the relevant impact events.

Certification Concept

A possible testing concept supporting fatigue and damage tolerance demonstration and complying with the applicable rules defined by the authorities within the certification specification document can be:

(1) The CFRP part is manufactured with a specially qualified material (Chap. 3) and a specially qualified process (Chap. 4). The maximum allowable manufacturing defects (porosities, delamination, etc.) and the maximum allowable (non-visible) impact damages are artificially introduced to the CFRP part.
(2) Dynamic loads are applied to the part, simulating a complete design service goal ("fatigue phase"). A load elevation factor is used and applied in order to compensate for any material strength reduction due to temperature and humidity. The factor depends on the type of material.
(3) The static ultimate load is applied, and the part must not fail.
(4) Visible and large impact damages are introduced to the part.

(5) Dynamic loads are applied for another design service goal (or a portion), and for this "damage tolerance phase" a load elevation factor is applied again to compensate for any material strength reduction due to temperature and humidity.

(6) A final test is made to determine the residual strength of the part, which should be above limit load.

Questions

1. How long does the aircraft development process typically take from the definition of top level requirements and the concept phase until the delivery of the first aircraft to the customer?
2. Why it is important to evaluate requirements (and possible solutions) of a new aircraft project at the beginning? What exactly needs to be assessed?
3. Which three parties play the most important role for requirements and the specification of a new aircraft?
4. What are typical general requirements for the development of commercial aircraft concerning manufacturers, operators and authorities?
5. Why are requirements from manufacturers, operators and authorities often conflicting? Examples?
6. Into which phases can the design process in accordance with VDI 2221 be divided?
7. During which phase of an aircraft product and component development would perform a functional analysis? Why?
8. In which phase of the development process of a new aircraft would you trade different materials? Why?
9. What is a functional analysis?
10. How is a functional analysis performed?
11. What is the benefit of a functional analysis?
12. What is a typical "design service goal" of a commercial aircraft?
13. Which typical structure loads do you know?
14. Where can you find certification relevant requirements for the aircraft structure ?
15. What are the typical phases of the certification process?
16. How do you principally proof, that the structure meets all safety and certification requirements?
17. How can you provide structural verification?
18. What advantage does the so-called pyramid test provide, and how it is structured?
19. What is "Limit Load", what is "Ultimate Load"?
20. What is a structure safety factor?
21. What is a structure reserve factor?

22. Why and how did the safety requirements for commercial aircraft increase over the past decades?
23. Prepare a simplified flow chart with the main steps of structure stressing and certification.
24. Is it reasonable to assume that load carrying CFRP structures which were damaged by impact show crack growth during aircraft operation?
25. Which typical impact damages must be considered when designing CFRP structures?

Exercise: Simplified Functional Analysis

1. Prepare a bubble diagram (see Fig. 2.6) to identify important functions of a passenger floor structure. Put the floor structure bubble in the centre and include all relevant elements in surrounding bubbles that might interfere with this structure throughout the complete lifetime. Distinguish at least the following phases: Assembly, Operation, Maintenance, Recycling.
2. Define the most important functions by reflecting the interfaces between the passenger floor structure and the interfering elements.
3. Define and describe the functions more precisely by means of a table with the following six columns: Phase (for all the phases stated in (1), Function, Criteria, Level (try to quantify each criterion), Flexibility ("must have" or "nice to have"), Means of Control (planned validation and verification). If a distinct criterion cannot be attributed to a single function, the function must be split further into its sub-functions.
4. Try to identify the most cost and weight driving functions, and provide some rationale.

References

1. VDI 2221 Methodik zum Entwickeln und Konstruieren technischer Systeme und Produkte. Beuth Verlag
2. Feldhusen, J., Grote, K.-H. (eds.): Pahl/Beitz Konstruktionslehre. Springer, Berlin (2013)
3. Certification Specifications for Large Aeroplanes CS-25. www.easa.europa.eu
4. Federal Aviation Regulations FAR Part 25 – Airworthiness Standards: Transport Category Airplanes. www.faa.gov
5. Swift, T.: Fail-Safe Design Requirements and Features, Regulatory Requirements. American Institute of Aeronautics and Astronautics, AIAA 2003-2783, AIAA/ICAS International Air and Space Symposium and Exposition: The Next 100 Y, 14–17 July 2003, Dayton, OH
6. Wikipedia: Aloha-Airlines-Flug 243
7. Handbuch Strukturberechnung HSB, Ausgabe 24.03.2006; Blatt 55105-01 A 1986
8. Hachenberg, D.: Strukturmechanische Anforderungen und Randbedingungen bei der Gestaltung eines CFK-Rumpfes für den Airbus der nächsten Generation, DGLR Jahrbuch 2001, DGLR-2001-133

 9. Davis, G.W., Sakata, I.F.: Design Considerations for Composite Fuselage Structure of Commercial Transport Aircraft. NASA Contractor Report 159296, March 1981
10. Fualdes, C.: Airbus – composite@airbus, damage tolerance methodology. Presented at the FAA Composite Damage Tolerance & Maintenance Workshop, Chicago, 19–21 July 2006. https://www.niar.wichita.edu/niarworkshops/Workshops/CompositeMaintenanceWorkshop, July2006,Chicago/tabid/99/

Chapter 3
Material Technology

Abstract This chapter starts with a description of the most important material selection criteria for primary load carrying airframe structures. In addition, crucial material related risks are highlighted. Since it is of interest for many material trades and the down selection procedure, fundamental properties of common and new aluminium alloys, including fracture toughness and fatigue, are explained. After a short description of fibre metal laminates (FML) and titanium properties, basic characteristics of thermoset (epoxy) and thermoplastic (PEEK and PPS) resin are introduced. The manufacturing process of carbon fibres and their properties are highlighted. The description of the thermoset prepreg manufacturing technology is followed by the discussion of the most important laminate properties. Finally, different relevant aspects, the targets and the procedure of material qualification are explained.

Keywords Material selection criteria • Material trade • Material risks • Aluminium alloy • 2024 • 2524 • Aluminium-lithium • Aluminium-magnesium-scandium • Fatigue crack growth • Fracture toughness • Fibre metal laminate • FML • Glare® • Titanium • Epoxy resin • PEEK • PPS • Carbon fibre • Thermoset prepreg • Storage life • Tack life • Work life • Laminate property • Tensile strength • Compression strength • Shear strength • Glass transition temperature • Water uptake • Media resistance • Material qualification • Screening • Open hole strength • Filled hole strength • Surface hardness • Dual sourcing

Selection Criteria

There are various selection criteria for airframe materials and it is important not to consider material properties individually, but always in combination with the specific design solution. This design solution must adequately reflect manufacturing technology constraints.

However, as detailed design solutions are not yet available during the early development phases, the following features can give some guidance for a preliminary material selection or material trades:

- Low specific weight (low density)
- High strength (compression, tensile, plain and notched, bearing)

- High stiffness (E-modulus, shear modulus)
- Excellent damage tolerance behaviour (fracture toughness, crack growth)
- Excellent fatigue behaviour
- Low spread of properties
- Superior corrosion behaviour and media resistance (water, seawater, phosphate ester based hydraulic fluid with additives, kerosene, de-icing liquids such as ethylene or propylene glycol, aggressive cleaning agents such as methyl ethyl ketone, paint strippers...)
- High temperature stability
- Good fire behaviour (burn through behaviour, smoke, toxicity)
- Sufficient electric conductivity
- Easy to process, low manufacturing effort
- Low environmental impact (carbon footprint, toxicity, waste, recycling)
- Easy to repair (cosmetic repair and structural repair)
- High availability
- Low cost

Provided that the necessary manufacturing technology can be made available and that the material can be qualified in time, cost and weight are clearly the dominating criteria for the material selection of a part or even a complete component. The weight benefit of material A compared to material B has to be compared with potential cost disadvantages and this cost has to include manufacturing effort deltas. The additional product value for each kg saved is about $1000, depending on the type of aircraft and the fuel price. Manufacturing cost deltas (i.e. more cost accepted due to weight benefits) must not outbalance operational cost benefits. On the contrary, it should be considered that the operational benefit must somehow be shared between seller and buyer. In addition, for comprehensive material comparisons and a final material selection, it is necessary to also analyse repair and maintenance cost deltas in service and to perform life cycle analysis in order to evaluate ecologic impacts. Finally, all risks associated with a new (not yet qualified) material must be individually assessed with their probability of occurrence and their severity, i.e. their impact on weight, cost or schedule, and mitigation plans have to be established. Potential risks of new materials can be linked to (examples):

- Material crashworthiness
- Material resistance to high energy impacts such as bird strike, tyre burst, etc.
- Material resistance to in-flight fire or post-crash fire
- Thermal behaviour
- Acoustic properties
- Large damage capability
- Electrical system installation (see Chap. 9)
- Lightning strikes
- Corrosion prevention
- Assembly tolerances
- Surface quality (roughness, optical appearance, painting, paint stripping)
- Repair
- Other

In case new manufacturing processes are required, all risks linked to manufacturing must be assessed as well.

Whenever CFRP is the matter of investigation for airframe structures, comparisons to a well-known, qualified and established material as well as to the latest aluminium alloys and fibre metal laminates are required. If possible, investigations should be based on a fully sized part, and the design should reflect manufacturing constraints as appropriate for the relevant material. Direct comparisons of material prices in €/kg or $/kg are not useful, as not only the part weight but also the buy to fly ratio can be completely different for materials, depending on the manufacturing technology.

Conventional and New Aluminium Alloys

Figure 3.1 shows material families of aluminium alloys. Age-hardening alloys are typically used for skins and panels. The A320 and the B737 fuselage consist of 2xxx (Al-Cu-Mg alloy) aluminium sheet. The properties shown in Table 3.1 for 2024 are similar for 2524, but the latter material has an even better damage tolerance and higher residual strength.

Lithium and Magnesium have very low material densities of 0.534 g/cm^3 and 1.738 g/cm^3, respectively. New aluminium alloys with these materials show superior mechanical property to density relationships. Aluminium-Lithium was selected for the fuselage of the new Bombardier C-series [1].

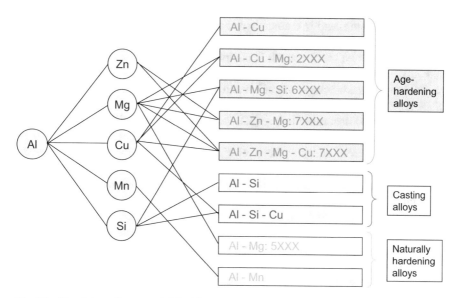

Fig. 3.1 Aluminium alloy material families

Table 3.1 Mechanical properties of aluminium alloys

Material	Tensile strength [Mpa]	Density [g/cm^3]	Yield strength [MPa]	Strain [%]	E-modulus [MPa]
1424	430	2.52	280	7	77,000
2024	420	2.78	310	15	69,000
2524[a]	420	2.78	310	15	69,000
6013	360–370	2.71	315–410	8–10	70,000
6056	360–370	2.71	315–410	8–10	70,000
AlMgSc	370–405	2.65	290–345	12–14	73,000

[a]Higher damage tolerance than 2024, higher residual strength

Aluminium-Magnesium-Scandium also offers interesting properties at reduced density (compared to 2024 and 2524). Both laser beam welding and friction stir welding have successfully been demonstrated in technology programs, and in combination with creep forming of flat, laser beam welded stringer stiffened panels into curved panels [2].

Fatigue Crack Growth

In Fig. 3.2, *L* means longitudinal (sheet production or rolling direction) and *LT* or *T* means the transversal direction. For the two letters *T-L* or *L-T* assigned to the individual material properties, the first letter is indicating the loading direction of the samples, and the second letter indicates the crack growth direction. If *a* is the crack length in mm, and *N* the number of dynamic load cycles, a crack of a sheet which is dynamically loaded with a stress $\Delta\sigma$ will grow as a function of *N*. The test samples to examine crack growth behaviour are typically sized 400 mm in width and 700 mm in length.

To compare the crack growth behaviour of different alloys, the derivative of the crack growth rate *da/dN* is plotted versus ΔK, the stress intensity factor. This factor can be calculated according to Eq. (3.1), where $\Delta\sigma$ is the applied stress delta, and *a* is the crack length. The comparison between the materials is then drawn by evaluating the stress intensity factor at a given crack growth rate. In case of the "Fatigue Crack Growth T-L" value shown in Fig. 3.2, the crack growth rate was set to 5×10^{-3} mm/cycle. The stress intensity factor at this growth rate of Aluminium-Lithium and Aluminium-Scandium alloys is higher than that of the reference material 2524; this means that higher cyclic stress $\Delta\sigma$ would have to be applied in order to reach the same crack growth rate as for the conventional aluminium alloy 2524. This is a clear benefit for fatigue crack growth sized parts, as for a given cyclic loading the wall thicknesses of Aluminium-Lithium and Aluminium-Scandium parts can be reduced compared to 2524 parts.

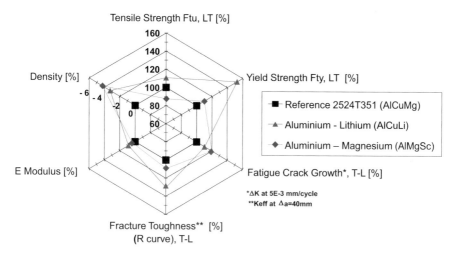

Fig. 3.2 Aluminium-Lithium and Aluminium-Scandium (properties compared to the reference material 2524 (AlCuMg) [2])

Fracture Toughness (Crack Resistance Curve or R-Curve)

A static test is performed to determine the crack resistance curve of a metal sheet material that has been dynamically loaded in advance to generate a fatigue crack. The tested specimen, with an artificial notch and a certain fatigue crack length a, the result of the cyclic loading, is statically loaded. The plot of the effective stress intensity factor K_{eff} versus the crack growth or extension (the crack will grow during the test) Δa is the R-Curve. K_{eff} can be calculated according to Eq. (3.2), where L is the load, A is the cross section of the test specimen, and a the crack length. During the R-Curve-testing, typically performed with test panels sized 700 mm in width and 1200 mm in length, the effective stress intensity factor K_{eff} will rise continuously to a maximum as the crack grows. In order to compare the fracture toughness of different aluminium alloys, the values measured for K_{eff} are compared for a defined crack extension Δa. In case of the fracture toughness or K_{eff} values presented in Fig. 3.2, Δa was set to 40 mm and the crack was running in longitudinal L direction where the part was loaded in transverse direction T (T-L). A higher K_{eff} value of Aluminium-Lithium and Aluminium-Scandium means an advantage over the conventional aluminium alloy 2524, as higher stresses would be necessary to generate the same certain crack extension or a complete failure (higher residual strength) of a structure component that is already weakened by a crack of a critical length.

$$\Delta K = \Delta \sigma \sqrt{\pi a} \qquad (3.1)$$

$$K_{eff} = \frac{L}{A}\sqrt{\pi a} \tag{3.2}$$

where

ΔK is the stress intensity factor [N/mm$^2 \cdot$ mm$^{1/2}$]
$\Delta\sigma$ is the applied stress delta [N/mm^2]
a is the crack length [mm]
L is the load [N]
A is the cross section of the test specimen [mm^2]

Finally, it must be pointed out that the pure density advantage of Aluminium-Lithium and Aluminium-Scandium compared to conventional aluminium alloy 2024 or 2524 is important as well. If, for example, the total fuselage panel weight for a new aircraft is at about 3 t, a pure density advantage (neglecting mechanical performance and other aspects of an improved design) of 4 % means a weight saving of 120 kg, which can make all the difference during the development phase of an aircraft.

The cost of conventional 2024 sheet material is at about 10 €/kg for standard geometries (the standard width is 2.5 m), but it can easily double if extra-long or extra-wide sheet material is needed. The cost of higher performing alloys such as Aluminium-Lithium and Aluminium-Scandium is higher, and therefore it has to be analysed case by case if the weight savings can outweigh the cost delta.

Special precautions are necessary if aluminium parts are attached to CFRP parts in order to prevent galvanic corrosion. Insulation layers of electrically non-conductive glass prepreg can be used for this purpose, see Chap. 7.

Fibre Metal Laminates

One of the best known and most commonly used fibre metal laminate is GLARE®. It is a hybrid material built up from alternating layers of aluminium and glass fibre reinforced epoxy prepreg, Fig. 3.3. Different types of GLARE® have been developed and investigated. For applications in civil aircraft, the aluminium sheet layer thickness is typically between 0.3 and 0.4 mm, and the aluminium material is 2xxx or 7xxx. The glass fibre prepreg thickness is typically 0.125 mm cpt (cured ply thickness) and applied in 0° and 90° direction. Different stacking sequences have been qualified for different airframe parts. The thickness of the basic hybrid sheet

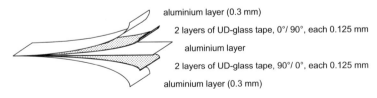

aluminium layer (0.3 mm)
2 layers of UD-glass tape, 0°/ 90°, each 0.125 mm
aluminium layer
2 layers of UD-glass tape, 90°/ 0°, each 0.125 mm
aluminium layer (0.3 mm)

Fig. 3.3 GLARE® laminate structure (example)

Fig. 3.4 Crack stopper function of glass fibres

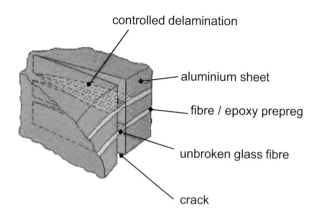

materials is usually between 0.85 and 1.95 mm. Furthermore, different epoxy prepregs have been qualified for 125 °C (in combination with the 2xxx alloy) cure and 180 °C cure (in combination with the 7xxx alloy), respectively. The advantage of the latter alloy is a higher maximum service temperature.

The main advantage over conventional aluminium sheet material is an improved fatigue/crack propagation behaviour. The glass fibres work as crack stoppers, Fig. 3.4.

Additional advantages are the relatively low density of 2.46–2.52 g/cm^3 (compared to conventional 2024 aluminium sheet with 2.78 g/cm^3; the density also depends on the exact type of GLARE®), the impact behaviour and the behaviour when exposed to fire.

On the downside, the manufacturing effort is relatively high. All aluminium foils need a chemical surface treatment, manual work is necessary for the placement of the foils in the tools, and complex part shapes can require expensive additional process steps in order to avoid unwanted wrinkling or other deficiencies. The curing of the epoxy prepreg, leading to a strong adhesive bond with the aluminium foils, requires the application of pressure and temperature (similar to CFRP) and expensive auxiliary material for bagging the parts. Depending on the part, the price can be up to 10 times higher compared to conventional 2024 aluminium sheet.

It also has to be pointed out that the material stiffness is lower (between 48 and 62 GPa, depending on the exact type of GLARE®) compared to conventional 2024 aluminium sheet (70 GPa). As the global load distribution within the airframe tends to follow the stiffness distribution ("stiffness attracts load"), this can be a disadvantage for a mixed material design with GLARE® and aluminium parts.

For A380, GLARE® was used for large parts of the fuselage structure as well as for leading aerofoil edges (Fig. 3.5).

Latest developments target the automation of the manufacturing process (manufacturing cost savings) and new types of GLARE®, using advanced aluminium alloys for the foils instead of conventional alloys to capture weight savings due to superior material properties.

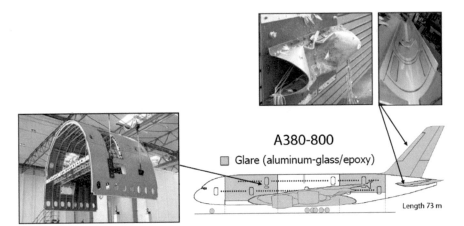

Fig. 3.5 Fuselage skin panels and tail plane leading edges made with GLARE® [3]

Another group of fibre metal laminates (FML) is the combination of titanium foils with CFRP. It has successfully been demonstrated that especially good pin loaded bearing strength properties can be achieved, important for the design of bolted structures [4]. However, until today this hybrid material is not the first choice for panel or skin applications in civil aircraft. Except for the relatively high material cost—mainly as a consequence of the manufacturing process—a key reason is that the implementation of an economic and robust process for the surface pre-treatment of the titanium foils prior to bonding is extremely difficult.

New attempts to combine advantageous properties of metals with those of CFRP are based on metal filaments and presently matter of R&D, see Chap. 9.

Titanium

Titanium is applied in many highly loaded and complex shaped airframe applications. Typical examples are:

- Fittings and brackets (for pylon, flap track, landing gear, nacelles, ...)
- Bulkheads (especially in temperature loaded areas)
- Frames
- Seat rails
- Roller tracks

A qualified titanium alloy is Ti-6 Al-4 V (alloy with 6 % aluminium and 4 % vanadium). It has a density of 4.4 g/cm^3, a tensile strength above 900 MPa and a Young's modulus of 115 GPa. Titanium parts can easily be combined with CFRP parts, as the dissolution potential between these materials is relatively low and no special precautions (insulating layers etc.) are necessary to prevent galvanic corrosion. The resistance to all relevant media (skydrol hydraulic fluid, kerosene, cleaning agents…) is excellent.

Titanium is, however, relatively expensive, and depending on the exact type of titanium alloy material cost can exceed 70 €/kg. Forged parts that need subsequent mechanical milling in dedicated areas are expensive, too. The typical milling volume is usually only 0.1 dm^3/min, while 4 dm^3/min can be achieved for aluminium parts.

Composite Matrix Systems

Epoxy Resin

More than 90 % of all primary load carrying composite components in airframe are based on epoxy resins. The chemical reaction mechanism is shown in Fig. 3.6. Epoxy thermosets qualified for airframe applications are non-meltable and (although water absorbing between 1 and 5 %, depending on the exact resin formulation and the conditioning) very resistant against media such as water, seawater, skydrol (a phosphate ester based hydraulic fluid with additives), kerosene, de-icing liquids such as ethylene or propylene glycol, and aggressive cleaning agents such as methyl ethyl ketone (MEK). The impact of these media on the final material properties must, however, be carefully analysed and is a matter of extensive test programs during material qualification.

The fracture behaviour of cured epoxy resin is brittle. However, epoxy resins used for airframe applications can contain additional materials such as thermoplastic or elastomeric polymers in certain quantities, improving the failure strain and material ductility [5]. The resistance against ultraviolet radiation is limited and paint is required for permanent protection. The temperature stability depends on the

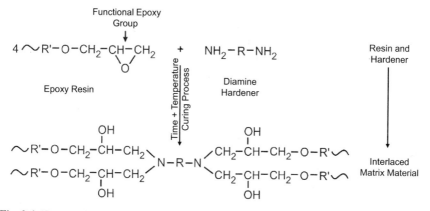

Fig. 3.6 Epoxy polyaddition reaction

Table 3.2 Typical properties of an epoxy thermoset resin

Tensile strength	MPa	60–80
Tension-E-modulus	MPa	3000–4000
Failure strain	%	2–8
Density	g/cm³	1.1–1.3
Shrinkage	%	1–3

degree of cure and the degree of cross linking (interlacing) within the molecular structure.

Mainly two systems are qualified and widely applied for airframes: a 125 °C curing system and a 180 °C curing system, the latter being more temperature stable than the first one, depending on the exact type of material up to approx. 120 °C under wet conditions. The maximum service temperature is limited by the glass transition temperature T_G of the cured epoxy resin and a safety margin, typically a delta of 30 K. For this purpose, the glass transition temperature T_G should be determined for conditioned material, i.e. after the material specimen has been exposed to humidity and temperature. This is usually achieved within an environment of 85 % rel. humidity at a temperature of 70 °C until equilibrium is reached for the water uptake. The glass transition temperature T_G of conditioned epoxy resins can be lower than that of dry resins. Some important material properties of epoxy resin are shown in Table 3.2.

Thermoplastic Matrices

Only a few % of all primary load carrying composite components in airframe are based on thermoplastic resins. The main thermoplastic matrix systems qualified for airframe applications are PEEK and PPS.

PEEK is a semi-crystalline polymer with a density of 1.32 g/cm³ and a melting range between 340 and 380 °C. The glass transition temperature T_G is reached at approx. 140 °C. The tensile strength is 90–100 MPa, and Young's modulus is 4.0 GPa. The elongation at break is between 40 and 50 %. PEEK is very resistant against all media relevant for airframe structures. The water absorption is low (approx. 0.5 % at saturation when exposed to water at room temperature).

PPS is a semi-crystalline polymer with a density of 1.34 g/cm³ and a melting range between 280 and 290 °C. The glass transition temperature T_G is reached at approx. 90 °C. The mechanical properties of carbon fibre reinforced PPS are below those of carbon fibre reinforced PEEK. The tensile strength of pure PPS is below 85 MPa, and Young's modulus is 3.9 GPa. The elongation at break is approx. 20 %. PPS is very media resistant. The water absorption is low (below 0.1 % at saturation when exposed to water at room temperature). Table 3.3 shows a comparison of PEEK, PPS and toughened epoxy resin properties.

Compared to epoxy based material, carbon fibre reinforced thermoplastics demonstrate an excellent hot/wet performance of strength properties relevant for

Table 3.3 Comparison of PEEK, PPS and toughened epoxy properties

Property	Unit	PEEK	PPS	Epoxy (toughened)
Density	g/cm³	1.32	1.34	1.3
Melting range	°C	340...380	280...290	n/a
Glass transition temperature (dry)	°C	140	90	150...170
Tensile strength	MPa	90...100	75...85	60...80
Youngs modulus	GPa	4	4	3...4
Elongation at break	%	40...50	5...10	2...8
Max. water absorption at room temperature	%	~0.5	~0.05	1...5

light weight design, Table 3.4. In addition, their in plane shear strength and their fracture toughness (G_{ic}) are excellent. Even above their glass transition temperature semi-crystalline thermoplastics such as PEEK and PPS maintain a relatively high level of stiffness and strength, since only their amorphous regions become rubbery-elastic.

Thermoplastics are meltable. Forming and welding is possible above the melting range. Due to the stamp forming technology, carbon fibre reinforced thermoplastic composites are very interesting for small parts in large quantities, such as connecting elements (clips, cleats, brackets, [7, 8]). However, as material prices are relatively high (the price a carbon fibre reinforced UD PEEK prepreg is approx. 100–200 €/kg, depending on the type of fibre) and as an efficient manufacturing technology enabling large components (wing covers, fuselage skins) is not available, the application for airframe structures today is rather limited, see Chap. 4. Present R&D is focussing on improved processing technologies for rapid thermoplastic tape placement processes [9] as well as on low cost and high performance polymer blends [10] and improved electrical properties [11].

Other Polymer Matrices

Phenolic resins are used for interior cabin applications, especially due to their excellent FST behaviour (flammability, smoke, toxicity), but not favourable for primary load carrying structures, as the mechanical properties are worse than those of epoxy, PEEK or PPS.

Polyester resins are also not preferred for airframe applications due to their poor performance under hot/wet conditions.

Table 3.4 Comparison of strength properties between epoxy and PEEK laminates, all data provided by TohoTenax [6]

Test	Orientation	Cond./ Temp.	Method	Unit	Tenax®-E TPUD PEEK-HTS45	Tenax®-E HTS40 F13 12K Epoxy (Aero)
Tensile modulus	0°	RT	EN2561	GPa	142	144
Tensile strength	0°	RT	EN2561	MPa	2450	2254
Compression modulus	0°	RT	EN2850-A4	GPa	130	121
Compression strenght	0°	RT	EN2850-A4	MPa	1578	1468
		HW/ 70 °C	Property loss		2–3 %	10–30 %
In-plane shear modulus	±45°	RT	EN6031	GPa	5.5	4.8
In-plane shear strength	±45°	RT	EN6031	MPa	144	104

Matrix, fiber-matrix dominated: Improvement vs. CF/EP (ambient temp.); large improvements vs. CF/EP (hot/wet)
Fiber dominated: Equivalent to CF/EP (ambient temp.)

Carbon Fibres

The dominating fibre material within primary load carrying airframe structures is carbon. Aramid fibres are sensitive to moisture absorption and only used for very specific applications. Glass fibres can provide interesting strength properties, but even though they are cost attractive, their stiffness is 3–4 times lower than that of carbon fibres and their density is higher, limiting their light weight potential, and only few parts of modern civil aircraft are made with glass fibre reinforcements. The following subchapter will focus on carbon fibres.

The history of carbon fibres for technical applications goes back to the nineteenth century. Key milestones were according to [12]:

- Patents of Swan (1878) and Edison (1879) for glowing filaments for lamps
- 1940...1950 basic research with PAN precursor at DuPont and Union Carbide
- 1960...1970 Shindo and Watt progress to control the shrinkage of PAN
- 1971 first commercial carbon fibre production at Toray, Japan
- 1986 first carbon fibre production plant in Germany (Tenax Fibres, Oberbruch)

There are different forms of carbon fibre appearance, Fig. 3.7, and different forms of carbon fibres in different scales, Fig. 3.8. Carbon fibres used for airframe applications have a diameter of 5–7 μm and their inner architecture is based on the graphite structure, Fig. 3.9. A single carbon fibre as shown in the SEM-image in Fig. 3.9a contains a lamellar structure with several adjacent graphite layers. A

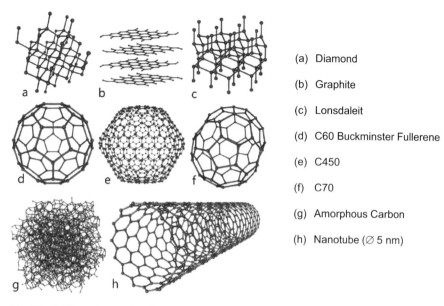

(a) Diamond

(b) Graphite

(c) Lonsdaleit

(d) C60 Buckminster Fullerene

(e) C450

(f) C70

(g) Amorphous Carbon

(h) Nanotube ($\varnothing$ 5 nm)

Fig. 3.7 Different forms of carbon appearance [13]

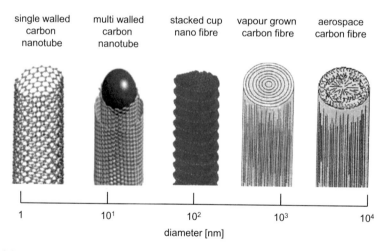

Fig. 3.8 Carbon fibres in different scales

TEM-image with these graphite layers is shown in Fig. 3.9b. For a better perception, these graphite layers are schematically depicted in Fig. 3.9c. Within the graphite layers, the carbon atoms form hexagonal honeycombs. A schematic magnification is shown in Fig. 3.9d. In the transversal direction, the van der Waals bonds between the adjacent graphite layers are relatively weak. The high

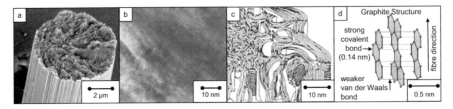

Fig. 3.9 Carbon fibres and typical inner graphite architecture

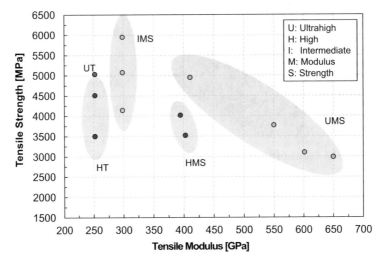

Fig. 3.10 Characteristics of commercially available carbon fibres [12]

tensile strength of carbon fibres in fibre direction is due to the high strength of the covalent bonds between the carbon atoms within the hexagons, Fig. 3.9d.

Different types of carbon fibres are commercially available. A few examples are given in Fig. 3.10. Most commonly used for airframe applications are fibres with a tensile modulus of 240 GPa and a tensile strength of 3600 MPa ("HT") as well as fibres with an intermediate modulus of 295 GPa and a tensile strength of 4700 MPa ("IMS"). Fibres with higher modulus require longer temperature treatment under inert gas atmosphere during fibre manufacturing and are more expensive. The cost of qualified standard-modulus fibres for structural applications within airframes is between 40 and 60 €/kg.

Carbon fibres for airframe applications are usually based on polyacrylnitril (PAN) synthetic fibres. Common yarns ("rovings") consist of 1 k (a bundle of 1000 filaments), 3 k, 6 k, 12 k and 24 k filaments; the corresponding linear density is 67, 200, 400, 800 and 1600 tex (g/1000 m). Some important properties of carbon fibres in comparison to glass and aramid fibres can be found in Table 3.5.

The manufacturing process of PAN based carbon fibres is shown in Fig. 3.11. The PAN-fibres are wetted in order to avoid unwanted electrical charging and

Table 3.5 Comparison of fibre properties

Fibre		Density, ρ [g/cm^3]	Tensile strength, σ [N/mm^2]	Elastic modulus, E [kN/mm^2]		Failure strain, εB [%]	Coefficient of thermal expansion, α [10^{-6} K^{-1}]	
				E∥	E$^\perp$		α∥	$\alpha^\perp$
Glass	E	2.6	2400	76	73	3	5	5
	R	2.53	3500	86	86	4.1	4	4
Carbon fibre	HM1	1.96	1750	500	5.7	0.35	−1.5	15
	HM2	1.8	3000	300		1	−1.2	12
	HT	1.78	3600	240	15	1.5	−1	10
	HST	1.75	5000	240		2.1	−1	10
	IM	1.77	4700	295		1.6	−1.2	12
Aramid	HM	1.45	3000	130	5.4	2.1	−4	52
	LM	1.44	2800	65		4.3	−2	40

HM high modulus, *HT* high tensile strength, *HST* high failure strain, *IM* intermediate modulus, *LM* low modulus

stretched. In a first oxidative process, hydrogen is separated and the molecular structure of the fibres is oriented. The fibres become non-meltable, which is an important prerequisite for the subsequent carbonisation. This happens in an oven under inert conditions, for HT fibres typically at temperatures between 1200 and 1500 °C. Fibres with a higher modulus require a treatment at higher temperatures up to 3000 °C. As it can be seen in Fig. 3.9, the graphite structures within the carbon fibres are not perfect, and this is important for the strength of the fibres and the strength of the composites: Perfect graphite layers could enable unwanted relative movements to each other, and perfect graphite layers at the surface of the fibres would lead to poor fibre-matrix bonds. Unsaturated valences of carbon atoms at the fibre surface can react with polymers and create strong covalent bonds, significantly contributing to improved mechanical composite properties. After carbonisation, a surface treatment can be necessary before the fibre bundles are wound up on bobbins. The typical width of a bobbin is 250 mm and the length of a 12 k 800 tex roving on a bobbin is typically 2000 m. The surface treatment can ease subsequent textile manufacturing operations and avoid unwanted fibre breakages; it can also be used to improve the fibre-matrix bonding in the composite. Epoxy-based resins can be used for these pre-treatments if the fibres are to be processed to composites with epoxy resins, however, different treatments can be necessary for thermoplastic polymer composites [12].

The typical production yield is approx. 1 kg of carbon fibres for 2 kg of PAN fibres. The typical capacity of a carbon fibre plant is 1000 t/year. Oxidation and especially carbonisation is very energy intensive, and since inert gases (typically 3000 m^3/h nitrogen or even argon) are needed, the amount of energy required for 1 kg of carbon fibres can double that of 1 kg aluminium sheet, see Table 1.1. This also has to be taken into account for material trades and life cycle analysis of parts.

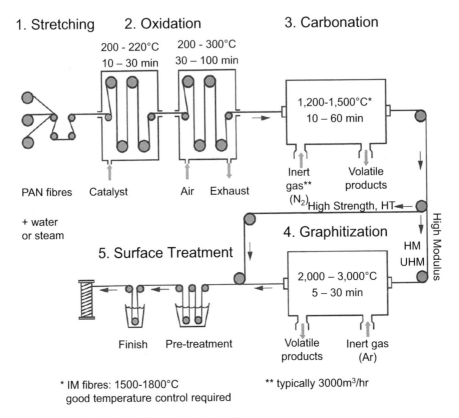

1. Stretching   2. Oxidation   3. Carbonation

Fig. 3.11 Manufacturing of PAN-based carbon fibres

Thermoset Prepregs

More than 70 % of all CFRP airframe parts in civil aircraft are based on thermoset prepregs. A prepreg (sometimes also "pre-preg" for *pre-impreg*nated) is a planar, semi-finished material which consists of fibres and resin. As epoxy resin is dominating the present applications, the following descriptions refer to these types of prepregs only. For thermoplastic prepreg processes see [9]. Most prepregs are produced with fibres lying parallel to each other in unidirectional orientation, but there are also prepregs made from fabrics. The resin content is typically 35–40 wt% (equalling 65–60 wt% fibre content), depending on the subsequent composite part manufacturing process. In case "zero bleed" processes are applied without any matrix losses, a 35 wt% matrix content of the prepreg will theoretically lead to ~60 vol% fibre content of the composite part. Due to inhomogeneity of semi-finished materials and manufacturing process tolerances, the fibre volume content of cured parts usually has a tolerance of 4 %. This must be taken into account for design allowables used for the stressing of composite parts. The fibre volume content tolerance also has impacts on part thickness tolerances and assembly.

Equation (3.3) can be used to convert the fibre volume content v_f to a fibre weight content m_f, and vice versa with Eq. (3.4), if ρ_f is the density of the fibre (typically 1.77 g/cm^3 for an HT-carbon fibre) and ρ_m is the density of the matrix (typically 1.3 g/cm^3 for a toughened epoxy).

$$v_f = \frac{1}{1 + \frac{\rho_f}{\rho_m} \frac{(1 - m_f)}{m_f}} \tag{3.3}$$

$$m_f = \frac{1}{1 + \frac{\rho_m}{\rho_f} \left(\frac{1}{v_f} - 1 \right)} \tag{3.4}$$

where

v_f is the fibre volume content [1]
m_f is the fibre mass content [1]
ρ_f is the density of the fibre [g/cm^3]
ρ_m is the density of the matrix [g/cm^3]

The thickness t (in mm) of a cured laminate with n_L layers can be calculated for "zero bleed" processes according to Eq. (3.5), where m_A is the prepreg areal weight of the single prepreg layer (in g/mm^2), ρ_f is the density of the fibres (note that it important to apply the unit g/mm^3 instead of g/cm^3). n_L can be set to 1 in case the thickness has to be calculated for a single layer. Youngs modulus E_c of a cured laminate with pure unidirectional fibre orientation can be calculated according to the rule of mixture applying Eq. (3.6), where v_f is the fibre volume content, E_f is the tensile modulus of the fibre, v_m is the matrix volume content and E_m is the tensile modulus of the matrix.

$$t = \frac{m_A n_L}{\rho_f v_f} \tag{3.5}$$

$$E_c = v_f E_f + v_m E_m \tag{3.6}$$

where

t is the laminate thickness [mm]
m_a is the single ply areal weight [g/mm^2]
n_L is the number of plies [1]
ρ_f is the density of the fibre [g/mm^3]
v_f is the fibre volume content [1]
E_c is Young's modulus of the composite [N/mm^2]
E_f is Young's modulus of the fibre [N/mm^2]
E_m is Young's modulus of the matrix [N/mm^2]

Prepregs are delivered on coils. The coil width depends on the manufacturing process technology; a typical coil prepreg width is 600 mm. 300 mm and even smaller widths (slit tape, down to ¼ in.) are also possible for fibre placement operations, see Chap. 4, but small tapes can be more expensive due to cost and

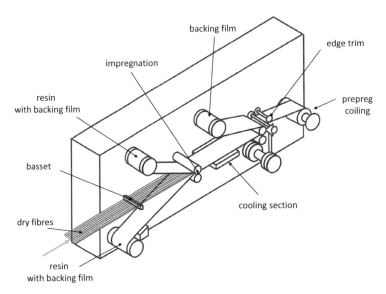

backing film

edge trim

impregnation

resin
with backing film

prepreg
coiling

basset

cooling section

dry fibres

resin
with backing film

Fig. 3.12 Prepreg production principle. Typical dimensions for a 1200 mm prepreg width production line are approx. 2 m in width and 20–30 m in length

waste of slitting processes. The prepreg manufacturing process principle is shown in Fig. 3.12.

Dry fibre rovings are pulled under a defined tension in parallel from several bobbins and spread as evenly as possible with a basset. However, this fibre distribution is far from perfect and causes inhomogeneity (approx. ±3 % in areal weight) and relatively high thickness tolerances. Gaps between rovings have to be avoided and should be well below 1 mm. The alignment of the fibres is also imperfect. Any deviations from 0° reduce mechanical properties, and the same applies for unwanted undulations. The number of fibre rovings (and their tex) depend on the desired fibre and prepreg areal weight. The higher the desired areal weight, the higher the tex and the number of rovings; however, a higher tex roving will be more difficult to spread than a lower tex roving.

Typical values for unidirectional prepreg tape areal weight range from approx. 200 g/m^2 to 400 g/m^2, and different grades of prepreg areal weights are qualified for different purposes. Cured ply thicknesses (cpt) of individual layers of 125 μm, 184 μm and 250 μm are used in many cases. Multi-axial laminates built up by thin individual unidirectional layers will tend to demonstrate superior mechanical properties compared to laminates of the same total thickness and the same share of 0°, ±45° and 90° layers, which built up by thicker individual prepreg plies. This is predominantly due to the fact that the amount of shear load which has to be transferred from one layer to the adjacent one with another orientation will rise with the ply thickness. The manufacturing cost benefit (due to higher lay down rates for higher areal weights) must therefore be traded against potential weight penalties (due to worse mechanical properties for high areal weights).

In a nip between two heated rollers the fibres are impregnated with resin, which is fed in from one or two coils (top and bottom) with a backing film, Fig. 3.12. In most cases, this is not a full micro-impregnation of all individual filaments, since dry areas between adjacent fibres are quite common. The micro-impregnation—free of voids—has to be insured by adequate subsequent processes during airframe part manufacturing. However, an even distribution of resin and an adequate level of impregnation should be ensured during prepreg manufacturing in order to avoid "dry spots", porosity or unwanted resin rich areas during later part manufacturing. The material cools in a cooling section before trimming the edges to the specified width and winding on a coil (mostly paperboard). A protection film (or backing film) is used to separate the prepreg layers on the coil and to enable easy handling of the prepreg during later manufacturing steps. The temperatures chosen for the impregnation at the nip point and the time of its effect on the epoxy resin have to be chosen carefully: The temperature has to be high enough to lower the resin viscosity and thus ensure a sufficient degree of fibre impregnation, while the chemical reaction leading to interlacing has to be limited at this stage. The material must maintain a certain tack and a certain formability for the manufacturing operations before its final cure. On the other hand, a certain degree of interlacing can help to stabilise the matrix for the granted storage and work life.

The impregnated fibres are wound up together with the protection film (backing film). In most cases this is a thin (40 μm) polyethylene film, separating the prepreg layers from each other, avoiding unwanted bonding of adjacent layers and enabling easy subsequent processing and handling. The typical length of a prepreg later used for automated tape laying processes (ATL) on a coil is 300 m.

The coils are bagged and stored at a temperature of $-18\,^{\circ}$C or below in order to slow down unwanted reaction (interlacing) of the epoxy matrix. The storage life usually granted by the prepreg manufacturers is 12 months at $-18\,^{\circ}$C. The work life (sometimes also called shop life) of an epoxy prepreg in the manufacturing shop of the airframe manufacturer is typically 20 days, i.e. 10 days tack life (for lay-up operations) plus an additional 10 days "stand by time" at normal work conditions (i.e. usually at a temperature of $25\pm2\,^{\circ}$C, and a relative humidity of $45\pm10\,\%$), Fig. 3.13. Prepregs must not be used after this period is elapsed, since the allowables (special material properties used for part design) generated for the qualification of the material will not account for impermissible pre-cure of prepregs.

Prices for prepregs range between 50 and 200 €/kg, depending on the types of fibres and resins, thickness grades, width and quantities.

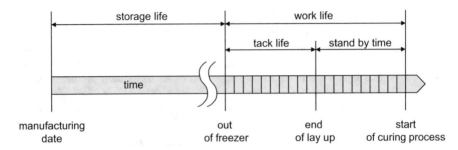

Fig. 3.13 Storage life, work life, tack life and stand by time of epoxy prepregs

Laminate Properties

Some typical stress properties of 180 °C cured laminates made with intermediate modulus (IM) carbon fibres at different conditions are shown in Table 3.6. Tensile strength and modulus are clearly fibre dominated; note that according to Eq. (3.6) the share of the matrix for the composite modulus of a unidirectional laminate can be neglected, as the stiffness of the matrix is approx. 70 times lower than that of the IM carbon fibre). The tensile stiffness is unaffected by moisture and temperature in this range, as in the pre-failure phase the matrix is almost irrelevant for the load transfer in a test sample with endless, unidirectional fibres. The tensile strength, however, is reduced at elevated temperatures and with moisture uptake. As the matrix softens under temperature and moisture influence it also loses strength, and the fibre matrix bonding can be negatively impacted, too. The lack of ability to transfer high shear loads between the fibres leads to reduced laminate tensile strength of approx. 15 % at 90 °C/wet compared to room temperature/dry conditions. For a bi-axial laminate with ±45° fibre orientation in relation to the loading direction, the matrix is dominating the properties. As no unidirectional, endless fibres exist in this type of laminate, the loads have to be transferred by the matrix. Temperature and moisture negatively affect strength and stiffness. The strength drop between room temperature/dry test conditions compared to 90 °C/wet is almost 25 %.

Material Qualification

The purposes of material qualification for primary airframe applications are manifold.

The main targets are (list not exhaustive)

- to define crucial material properties, assess (quantify) the relevant limit values, and verify and certify that these material properties can repeatedly be achieved with high reliability (i.e. with a defined, high level of confidence)

Table 3.6 CFRP properties of laminates with 60 % fibre volume content at room temperature, at 70 °C and at 90 °C (IM fibres, epoxy resin)

		RT/ dry	70 °C/ dry	70 °C/ wet	90 °C/wet[a]
Tensile strength (0° laminate)	[MPa]	2786	2585	2577	2354
Tensile modulus (0° laminate)	[GPa]	175	175	177	176
Shear strength (±45° laminate tensile)	[MPa]	112	107	85	84
Shear modulus (±45° laminate tensile)	[GPa]	4.4	4.1	3.6	3.4

Data: Courtesy of Cytec Engineered Materials Limited (Solvay SA)
[a]Wet = pre-conditioning in an atmosphere of 70 °C and 85 % relative humidity up to an equilibrium (water uptake <2 %)

- to provide a database for the generation of design allowables
- to provide a database for quality assurance of incoming material for production
- to provide a database for the decision process on concessions, repair or scrap for quality findings during or after part manufacturing
- to provide a database for repair decisions and repair methods
- to provide a database as a reference for decisions on second source or alternative materials

In case a new material is developed for a new aircraft, the qualification process should be fully completed for type certification, see Fig. 2.1.

This means that

- all relevant material properties are defined and quantified
- all tests are carried out and fully evaluated
- all relevant data is recorded
- all documents are available and signed

As final material properties of a composite are influenced by the manufacturing process, the existence of a fully defined, qualified manufacturing process is an important prerequisite for material qualification. Manufacturing process conditions must be recorded and quoted in material specifications.

The definition of *all* material properties crucial for its function as load carrying airframe that should be included in the material qualification process and its documentation is very important. Completeness should be insured by multi-disciplinary teams, involving experts from design, stress, materials, manufacturing, quality, system installation, assembly, surfaces, maintenance and repair, etc. Systematic methods as described for functional analysis in Chap. 2, reflecting the complete "life" of the material from cradle to grave can be helpful. Strictly formal, a material can also be regarded as "qualified" due to the existence of signed material specification documents if it has been carried out independently from a specific design of a given part. However, practise has shown that it can be very important to reflect specific part requirements already at the beginning of a new material qualification project, in order to ensure that all relevant material properties have been identified and quantified.

For a conservative and safe design, design allowables, i.e. material properties taken for the stressing of composite parts, must be derived from laminate properties which take worst case conditions into account. These conditions must include very low temperatures (usually $-55\,°C$, but in some cases, the material behaviour has to be investigated at even lower temperatures) as well as very high temperatures (70–$120\,°C$, depending on the surface finish, the type of paint and its colour, the position of the part within the aircraft and heat flux etc.). Heat flux calculations and temperature mappings can be necessary. Worst case conditions also include the impact of moisture (water uptake) and aggressive fluids, and the combination of these impacts with temperature. Furthermore, worst case conditions also mean that material data must cover the boundaries of tolerances of the semi-finished material (prepreg) such as fibre/resin content, fibre waviness, thickness, gaps, storage life, tack life, work life etc. In addition, the allowables must account for the tolerance borders of the manufacturing process (applied pressure, heating rates, curing time, cooling conditions), which also include maximum allowed porosity and maximum allowed deviation from the desired fibre orientation. As carbon fibre reinforced polymers are sensitive to impact, and impacts can happen and remain undetected (as no damage can be seen from the outside) already during manufacturing and assembly, data must also be collected for impacted specimens.

Finally, the data generated by the material tests must be statistically secured by a large number of test specimens (usually a minimum of 15) for each individual property, which must be made from different production batches. For the generation of allowables, reduced values are calculated (well below mean values), ensuring a high probability to reach or exceed these values when the material is used for production.

In summary, the following aspects of strength (such as tensile, compression, in plane shear, inter-laminar shear, out of plane G_{ic} and G_{iic}) and stiffness properties are essential for the qualification of a new CFRP material (minimum list, not exhaustive):

- cover the complete semi-finished material tolerances
- cover the complete manufacturing process tolerances
- include low and high temperatures
- include dry and wet conditions
- include notches (rivet holes)
- effect of impacts
- include media influence (water, skydrol, fuel, cleaning agents, paint strippers...)

In some cases the electrical properties of CFRP can also require the clear definition of design limit values and adequate testing (electrical conductivity, resistivity in-plane and out-of-plane etc.).

The assessment of notched strength properties is very important for the generation of design allowables. Due to holes required for rivets, the strength of composites can drop enormously. Notched properties must be investigated and assessed

Table 3.7 Typical CFRP laminate properties

		RT/dry	70 °C/wet[a]
Tensile strength	[MPa]	849	808
Open hole tension OHT	[MPa]	500	493
Compression strength	[MPa]	546	437
Open hole compression OHC	[MPa]	322	246
Filled hole compression FHC	[MPa]	426	391
Compression after impact (30 J)	[MPa]	227	213
Bearing	[MPa]	994	822

$v_f = 60\%$, IM fibres, epoxy matrix, $0°/\pm45°/90°$ laminate share of 25/50/25 (i.e. 25 % in 0°, 50 % in ±45°, 25 % in 90°)
Data: Courtesy of Cytec Engineered Materials Limited (Solvay SA)
[a]Wet: pre-conditioning in an atmosphere at 70 °C and 85 % rel. humidity to an equilibrium, afterwards tested at 70 °C temperature

while taking media and temperature impact into account. An example for a simplified method of design allowable generation is described in Chap. 6. Mechanical impacts (visible and non-visible from outside) can also have a significant influence on composite strength. Some notched composite properties are shown in Table 3.7. After conditioning (moisture and temperature impact), strength values drop between 5 and 24 % (exception: OHT) for material coupons tested at 70 °C when compared to the room temperature dry values. Notched specimens show strength drops of approx. 40 % when compared to un-notched (plain) specimens. For certain instances, as for the notched "open hole" tension specimens, the softening of the matrix due to elevated temperature and moisture uptake can positively influence the failure load, if some plastic deformation reduces local stress concentrations near the border of the notch.

For certain applications the surface hardness can be of interest, Table 3.8. Aluminium hardness values exceed those of CFRP, however, when the force is released the elastic effect of CFRP is very dominant. At a similar maximum load level of the same indenter, the remaining dent in CFRP is smaller than in aluminium.

Material Qualification Process

The material qualification process can be divided into different steps according to Fig. 3.14. A screening of material properties is necessary at the beginning in order to evaluate the potential for light weight design. Typical properties for a quick and easy assessment are (examples, non-exhaustive list):

- density
- glass transition temperature T_g
- tensile and compression strength, plain and notched

Table 3.8 Typical surface hardness (applied force per indenter contact area) of a CFRP laminate at room temperature

		CFRP	Aluminium
Martens	[MPa]	455	719
Vickers	[MPa]	54	76

HT fibres, epoxy matrix, $v_f = 60\%$, 50/40/10 laminate, dry, room temperature, Martens hardness measurement according to ISO 14577 1-3, Vickers ISO 6507-1

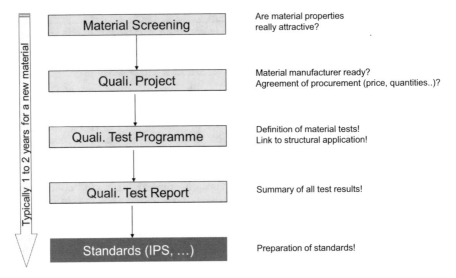

Fig. 3.14 Material qualification process

- modulus (tensile and compression)
- interlaminar shear strength ILSS

In case the material screening reveals positive results, a qualification project can be launched. For this purpose, several disciplines are mandatory. Engineering experts from different domains should be supported by persons in charge of manufacturing, quality and procurement. The time consuming and expensive qualification test program should only be initiated when clear agreements have been made with the material supplier (or, in case of dual sourcing, with the material suppliers), i.e. when negotiations on price, quantities, periods of supply, etc. were successful. Then a qualification test program has to be defined, which includes the definition of all relevant properties. The qualification test report is the documentation of all test results. Finally, the standards have to be generated.

Material standards can be defined for different types of materials such as:

- carbon fibre reinforced epoxy prepregs
- glass fibre reinforced epoxy prepregs
- glass/carbon hybrid reinforced epoxy prepregs

- carbon fibre reinforced thermoplastics
- resin transfer moulding liquid resins
- non crimp fabrics
- metallic prepreg mesh (lightning strike protection)
- honeycomb core
- structural adhesive films
- fillers
- ...(other)

For a given material standard, different specifications can exist, defining the material more specifically. For example, a material standard for "carbon fibre reinforced prepregs" could contain a specification for 180 °C curing tapes, and another specification for 180° curing fabrics.

These specifications should include at least:

- all important definitions of terms
- the description of the qualification process
- the description of the manufacturing process
- the maximum allowed manufacturing defects
- the number of samples and batches to be tested under certain conditions

The core of the material specification is the definition and assessment (quantification) of material performance requirements as well as for the semi-finished material (for example density, viscosity, areal weight, binder content...) as for the cured laminate (strength, modulus, T_g ...) under certain conditions (dry, media exposure, high and low temperatures...). All test specifications must be referenced. Material produced and delivered by the material manufacturer according to this specification must fulfil all these requirements. Individual material properties (tensile strength, for example) quantified in the material specification refer to the relevant test and manufacturing specifications for the test laminates. Minimum average values as well as minimum individual values can be specified.

Individual product properties, i.e. material properties of a certain product from a given material manufacturer, can be specified in individual documents which refer to the material specification. These documents can also be used for quality assurance purposes for incoming material in the manufacturing shop. Product properties may exceed limits defined in the relevant material specification, but not fall below.

This leads to the fact that a dual sourcing strategy—often favoured by procurement in order to obtain a better position for price negotiations with the material suppliers, and to improve the supply chain security for manufacturing—with two or more materials qualified to the same material specification will inevitably lead to lower individual property limits (always finding the lowest common denominator) in the material specification document, since two materials from different sources will not be completely identical in every aspect. In consequence, it is possible that a specific design solution is not fully weight optimised, i.e. not fully exploiting the potential of one material.

Questions

1. What are typical requirements for materials for structural components in aircraft?
2. Which are modern aluminium alloys that can be compared to composites in terms of performance and costs?
3. What is the density of aluminium aerospace alloys, what is the modulus of elasticity?
4. What is Glare®, and what are the advantages of this composite?
5. What is characteristic for the use of titanium?
6. How much weight reduction is typically possible with CFRP compared to Al in structural components?
7. Which resin system dominates the use of composites in aircraft?
8. Why is the titer (tex) of carbon fibre rovings used for the production of prepregs usually limited?
9. How are carbon fibres produced?
10. Why are carbon fibres strong?
11. What is the typical fibre content in mass% of prepregs?
12. How are prepregs produced?
13. What means "tack life" of a prepreg?
14. What is the Young's-modulus of standard fibres (HT)?
15. What is the diameter of carbon fibres in prepregs?
16. Do carbon fibres absorb water?
17. What are "hot/wet" conditions?
18. Can you name some aggressive media to which CFRP structures can be exposed during aircraft operation?
19. How strongly will the modulus of elasticity of a 0°-CFRP laminate (i.e. all fibres are parallel to the direction of stress) from room temperature/dry condition be reduced, when the test is carried out at 90 °C with samples that were previously conditioned at a temperature of 70 °C and 85 %-rel. humidity (until they reached an equilibrium state)?
20. How strongly will the modulus of elasticity of a ±45°-CFRP laminate from room temperature/dry condition be reduced, when the test is carried out at 90 °C with samples that were previously conditioned at a temperature of 70 °C and 85 %-rel. humidity (until they reached an equilibrium state)?
21. How strongly will the strength of a typical multiaxial CFRP laminate coupon be reduced if the coupon is notched (¼ in. hole)?
22. Name some different composite material systems used in aircrafts for structural parts.
23. Can you name some thermoplastic polymers used for airframe applications? What is their advantage over epoxy resin?
24. What is the purpose of material qualification?
25. What is described in a material specification?
26. What is described in an "Individual Product Specification"?

27. What happens during material qualification and how long does the qualification of a new material usually take?

Exercise: Second Source Material Qualification

1. The company "HPLCM" (High Performance Low Cost Materials") offers carbon fibre prepreg fabric as a candidate for a second source material of sandwich face sheets. The first source from a different material supplier is already qualified and used for series manufacturing. Prepare a planning for the qualification of the second source with all necessary steps/phases, define the main tasks within each phase and provide an estimate for the duration of all phases.
2. Which requirements (and related material properties) are the most important ones for the phases of design, manufacturing and operation?
3. For a first screening, which material properties would you investigate?
4. Assuming that the second source material candidate has lower, design relevant compression strength (-5%) compared to the already qualified first source, and all other relevant material properties are satisfying, i.e. meeting or even exceeding the level of performance as specified in the existing material specification. What is necessary in order to make the new second source material available for series manufacturing?
5. What kind of investigations would you propose to minimise the manufacturing risk for the second source material? Provide a checklist of all important quality aspects.

References

1. www.nowisthefuture.com
2. Knüwer, M., Tempus, G.: Development of High Performance Al-Alloys for Aircraft Application. Proceedings: Metals- The Competitive Edge, Institution of Mechanical Engineers, IMECHE, Sheffield, UK, 12 May 2009
3. Rans, C.D.: Evolution of FML Technology. TU Delft (2014)
4. Fink, A., Kolesnikov, B.: Hybrid titanium composite material improving composite structure coupling. ESA SP-581, European Space Agency, provided by the NASA Astrophysics Data System. European Conference on Spacecraft Structures, Materials & Mechanical Testing 2005, Noordwijk, The Netherlands, 10–12 May 2005
5. Friedrich, K., Breuer, U.: Multifunctionality of Polymer Composites: Challenges and New Solutions. Elsevier, June 2015
6. Roden, M.: Von der Kohlensstofffaser zum thermoplastischen Halbzeug, CCeV Thementag Thermoplaste, PAG, Bremen, 21 Juli 2015
7. Miaris, A., Edelmann, K., Sperling, S.: Thermoplastic Matrix Composites: Xtra Complex, Xtra Quick, Xtra Efficient-Manufacturing Advanced Composites for the A350 XWB and Beyond. Proceedings 20th International Conference on Composite Materials Copenhagen, 19–24 July 2015

8. Black, S.: Thermoplastic Composites "Clip" Time, Labor on Small But Crucial Parts. CompositesWorld, May 2015
9. Neitzel, M., Mitschang, P., Breuer, U.: Handbuch Verbundwerkstoffe. Carl Hanser Verlag, München (2014). ISBN 978-3-446-43696-1
10. Krooß, T., Gurka, M., Dück, V., Breuer, U.: Development of cost-effective thermoplastic composites for advanced airframe structures. Proceedings ICCM International Conference on Composite Materials, Copenhagen, 19–24 July 2015
11. Noll, A., Friedrich, K., Burkhart, T., Breuer U.: Effective multifunctionality of poly (p-phenylene sulfide) nanocomposites filled with different amounts of carbon nanotubes, graphite, and short carbon fibers. Polymer Compos. (2013). Published online 5.3.2013, doi: 10.1002/pc.22427
12. Wohlmann, B.: AVK Handbuch Faserverbundwerkstoffe. Vieweg + Teubner GWV Fachverlage GmbH, Wiesbaden (2010). ISBN 978-3-8348-0881-3
13. Koo, J.H.: Polymer Nanocomposites Processing, Characterization, and Applications. McGraw Hill Professional, New York, NY (2010). ISBN 0071492046, 9780071492041

Chapter 4
Manufacturing Technology

Abstract The selection of the manufacturing technology will inevitably influence cycle times and manufacturing cost, i.e. non-recurring cost as well as recurring cost. It will also influence quality assurance and assembly effort. Due to the strong link of advanced design schemes and suitable manufacturing technology, maintenance cost (for example with respect to the accessibility of integrated CFRP-structures) are also affected. Different manufacturing technologies require different semi-finished materials with different mechanical properties. Hence, the selection of the manufacturing technology will also influence part weight and operational cost. In this chapter, all state of the art as well as newly developed manufacturing technologies and recent R&D are described; advantages and shortfalls are highlighted. Automated tape laying, fibre placement and pultrusion technology are explained, and examples are provided. The autoclave process technology is discussed, including unwanted defects in the cured material and the description of quality assurance technologies (dielectric analysis and non-destructive testing by ultrasound). Different textile infusion technologies are characterised. Thermoplastic stamp forming technology and thermoplastic tape laying technology are discussed in detail. The description of the filament winding technology is followed by the presentation of joining and bonding technologies. Riveting as well as adhesive bonding and welding processes are discussed. The chapter closes with the description of a possible method for the selection of "the right process technology for the right part".

Keywords Prepreg • Autoclave • Automatic tape laying • Automatic fibre placement • Pultrusion • Floor beam • Skin-stringer integration • Wing skin • Co-bonding • Co-curing • Secondary bonding • Vacuum bagging • Curing cycle • Curing methods • Manufacturing defects • Material defects • Dielectric analysis • Ultrasound non-destructive testing • A-scan • B-scan • C-scan • D-scan • Textile infusion technology • Resin transfer moulding • Non-crimp fabric • Resin infusion • Resin film infusion • Resin viscosity • Rear pressure bulk head • Thermoplastic stamp forming • Press forming • Clip • Thermoplastic tape laying • Organo sheet • Shear deformation • Filament winding • Strut • Pressure vessel • Bolted joint • Hi-Lok® • Blind bolt • Riveting machine • Adhesive bonding • Adhesive film • Surface preparation • Peel ply • Water break test • Welding

© Springer International Publishing Switzerland 2016　　　　　73
U.P. Breuer, *Commercial Aircraft Composite Technology*,
DOI 10.1007/978-3-319-31918-6_4

Prepreg Autoclave Technology

The majority of composite airframe parts are made from prepregs by means of the autoclave technology. First, prepregs are taken out of the storage, i.e. a freezer with a temperature of -18 °C. In the second step, they are warmed up to room temperature and unpacked. Afterwards they are cut into shape, positioned in the desired lay up on a tool, vacuum bagged, evacuated, and exposed to temperature and pressure in autoclaves until the resin has cured. After cooling, the parts are unpacked, contour milled and quality checked.

During the last 20 years, manual prepreg cutting and placement processes have been more and more replaced by automatic processes; however, there is still a lot of manual work especially for bagging and debagging of parts. The cutting process is shown in Fig. 4.1.

The prepreg coils are stored in paternoster systems with computer supported monitoring of material types and their shelf life. Prepregs are positioned on a carrier and automatically cut into pieces by a static or an oscillating knife, which is guided on a portal and can move in x/y direction. Several thousands of individual patterns can be necessary for the different airframe parts.

The processing area is typically wide enough for prepreg coils of 1.8 m width, the length of the cutting table can be 30 m or more. Several layers placed on top of each other can be cut at once on a cutting table. Alternatively, it is possible to use a conveyer and to cut only one layer with a high cutting speed. Special software can be applied in order to optimise the arrangement of cut out patterns and to minimise production waste; however, waste cannot totally be avoided, Fig. 4.2. The production waste is usually cured, and a material "down cycling" is the consequence, i.e. shortened carbon fibres are reused for other products. The typical buy to fly ratio for the additive CFRP processing technology is between 1.2 and 2 (for comparison, for milled aluminium parts the ratio can be 20). Efforts are undertaken in minimising this ratio and developing new technologies in order to reuse production waste for airframes without material degradation in the future.

The prepreg cutting operations can be carried out in shops with or without air conditioning, but the limited shelf life of the prepreg (see Chap. 3) impacts the

Fig. 4.1 NC cutting of prepregs [1]

Fig. 4.2 Prepreg
production waste [1]

Fig. 4.3 Laser supported
positioning of prepreg
blanks [2]

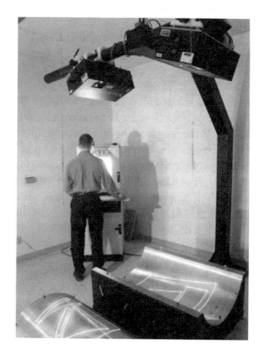

available time span for further process steps. Cutting should be performed in
contamination controlled areas.

The placement of prepreg blanks on the tools can be laser supported, Fig. 4.3.
Most tools larger than 3 m length are made of steel alloys with approx. 36 % nickel
(such as Dilaton or Invar), resulting in a very small thermal expansion coefficient
similar to that of CFRP. CFRP tools are also used, but their service life is limited
due to their wear sensitivity and surface erosion. For smaller parts, also steel tools
are used.

Automatic Tape Laying (ATL)

In order to position "endless" unidirectional prepreg layers on large tools, the automatic tape laying (ATL) technology is used. The technology became relevant for airframe series production during the 1980s, starting with planar lay ups for smaller parts, and has step by step been improved until today also for large and more complex shaped parts, Table 4.1. This is also due the fact that the pure size of the component, for example a wing skin, requires the use of automated equipment.

The tape laying head can be mounted on a portal, allowing for controlled movements in x and y as well as in z-direction, plus the necessary axis rotations to position the head vertically to the surface. The laying head carries the prepreg material. It is clamped and pulled by rollers, and laid down onto the tool in 0°, + and −45° and 90° direction, according to the laminate stacking sequence defined by design. A roll applies a defined pressure in z-direction in order to achieve the necessary adhesion of the tape on the layer underneath and to provide a pre-compaction of the laminate.

Modern ATL machines typically work with 300 mm tape width at maximum processing speeds of up to 300 m/h (equivalent to a deposition rate of approx. 15–20 kg/h, depending on the prepreg areal weight) and can handle typical curvatures and surface geometries needed for aerofoils (wing skins). However, deposition rates are reduced in case doublers (local thickenings) are necessary. The ramping of plies in order to cope with local thickness changes is possible in certain limits. As the length of the tapes on one roll is limited (to approx. 300 m) but several km of tape can be needed for a wing skin or a tail plane skin, the process has to be interrupted for material change.

The processing of functional materials such as glass prepreg (for corrosion prevention purposes in dedicated areas, see Chap. 7), peel ply or lightning strike protection is also possible. The machinery invest is in the order of 5 million euros, depending on functionalities and size. An example of a modern ATL machine is shown in Fig. 4.4.

Figure 4.5 shows a good example of an airframe structure made by ATL; it is the wing skin of A350.

Cutting waste cannot be completely avoided during ATL operations today. The existence of waste is evident particularly at the circumference of the parts; note the

Table 4.1 Development of ATL [3]

	1970s	1980s	1990s	2010s
Aircraft	A300	A310/A320	A330	A350XWB
Process	Manual layup	9-axis ATL	11-axis ATL	11-axis ATL
Complexity	–	2D—limited 3D	moderate 3D	3D with steering
Size		7 m	12 m	33 m
Material type	Woven prepreg	UD prepreg	UD prepreg	UD prepreg
Material width		75 mm	150 mm	300 mm
Functional materials	Manual layup	Manual layup	Manual layup	Automatic

Fig. 4.4 The Aeronautics TORRESLAYUP ATL machine is an 11 axes gantry CNC tape layer machine, that has been specially designed for high speed tape laying of compound contoured aircraft structural components [4]

Fig. 4.5 A350 stiffened wing skin (prepreg autoclave technology) [2]

"zigzag" at the edges of the laminate shown in Fig. 4.4. Waste can be reduced with smaller tapes; however, this compromises the deposition rate.

For complex shaped structures such as spars (u-shape) it is common to start with a flat ATL laminate, and to heat it by means of infrared radiation in a subsequent processing step. As the viscosity of the epoxy resin decreases with higher temperature, shaping of the lay-up is possible by means of an elastic diaphragm. For this purpose, the flat laminate is positioned on top of a male tool and heated. A diaphragm is used to cover the laminate and the tool, sealing the surrounding of the tool, and the air underneath the diaphragm is evacuated. The laminate is shaped by the delta pressure and cools down to the tool temperature, regaining sufficient rigidity for the following process steps (demoulding, handling, autoclave cure). For shapes with undercuts it can be necessary to apply a double diaphragm: The flat laminate is positioned in between an upper and a lower diaphragm, and the air between the diaphragms is evacuated. After heating, the laminate is formed into the desired shape by positioning the package above a tool and evacuating the air underneath the lower diaphragm. Double diaphragm processes can also help to avoid unwanted laminate wrinkling. The process is described in [5] and in [6].

Automatic Fibre Placement (AFP)

If complex shaped curved parts have to be manufactured, it can be necessary to use automatic fibre placement (AFP) processes, sometimes also referred to as "tow placement". A main difference to ATP is that separately controlled narrow tapes (normally 1/2, 1/4 or 1/8 in. width) are used, allowing to cover the tool surface with minimised unwanted gaps or overlaps even on non-geodetically paths, which would not be possible with "rigid" wide tapes used for ATP. However, the material can be more expensive than standard width prepreg, as it is manufactured by slitting prepreg ("slit tapes"), which is an additional costly step in the process chain.

Figure 4.6 shows an example of a modern AFP machine. The feed rate of the narrow tapes (up to 32 tows in parallel) can be controlled individually, and the tapes can be independently dispensed, clamped, cut and restarted during fibre placement. This allows a high flexibility for local thickness changes or cut-outs. In addition to the variance of the AFP head it is also possible to rotate the tool. In practise, the typical deposition rate is between 5 and 10 kg/h. Complex shaped fuselage panels can be manufactured by AFP.

A good example is the A380 rear end, Fig. 4.7, where a very complex geometry of the skins is necessary not only due to the taper of the fuselage, but also in order to provide the clearance for the ease-of-movement of the horizontal stabiliser (Fig. 4.7, left side, with an almost planar surface adjacent to the cut out).

Fig. 4.6 Fives Cincinnati viper fibre placement system [7]

Fig. 4.7 A380 rear end, fuselage skins made by AFP, frames by RTM [8]

Pultrusion

For long profiles with constant cross section, such as floor cross beams or skin stiffeners ("stringer"), automatic pultrusion can be an economic alternative to manual lay-up and forming processes. An automatic process for prepreg strips has successfully been developed by Jamco in Japan, Fig. 4.8. Prepreg strips with $0°$, $+$ and $-45°$ as well as with $90°$ fibre orientation are pulled from dispensers for a discontinuous pressing cycle. A contoured male and female tool is alternatingly opened during the feeding process, and closed again in order to apply heat and pressure. Final curing is achieved in a downstream oven.

T-shaped stringer profiles can easily be made by longitudinal cutting of H-shaped profiles. Lightening holes, as applied in the shear web of A380 floor cross beams, Fig. 4.9, require additional cutting processes. Automatic pultrusion is state of the art for constant cross sections and non-curved profiles, however, the process is much more difficult for complex shaped or curved stiffeners, such as fuselage frames, and requires further development.

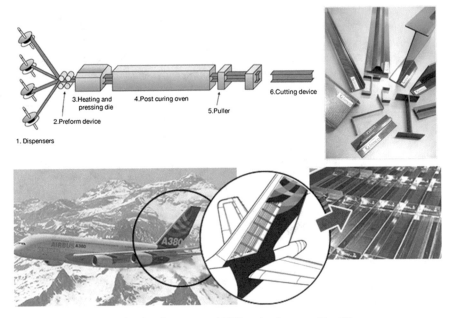

Fig. 4.8 Jamco advanced pultrusion process (ADP) and stringer profiles [9]

Fig. 4.9 A380 fuselage cross section with lower and upper floor structure (*left*) and CFRP upper floor crossbeam with lightning holes (*right*) [9]

Skin/Stringer Integration

Monolithic design schemes of load carrying airframe structures such as wing or fuselage panels will usually require stiffened skins in order to prevent unwanted buckling. Different technologies can be applied to integrate skin and stringer.

- *Co-Curing* means that several uncured parts are cured together in one step, creating an integrated part.
- *Co-Bonding* means that uncured parts are positioned on already cured parts (or vice versa). Heat and pressure is applied to cure the uncured parts *and* to bond them to the already cured parts at the same time.
- *Secondary Bonding* means that cured parts are integrated by adhesive bonding (usually by means of heat and pressure) in a separate process.

Details as well as advantages and disadvantages of different integration technologies are described in Chap. 8.

Autoclave Process

Autoclaves are commercially available for various part sizes and process parameters. The diameter of autoclaves used for airframe manufacturing can reach 8 m or even more, and a length of more than 35 m.

The temperature can reach more than 400 °C (especially for thermoplastic applications with PEEK) and the pressure more than 30 bar, but for most cases of thermoset epoxy prepreg curing 180 °C and 7 bar are used.

Nitrogen gas is applied for an inert atmosphere within the autoclave to avoid inflammability. Figure 4.10 shows a bagged prepreg lay-up ready for the application of heat and pressure within the autoclave. In most cases, the bagging is a manual process. A typical bagging scheme is shown in Fig. 4.11. The tool is

Fig. 4.10 Vacuum bagged prepreg part with vacuum hoses ready for the autoclave process

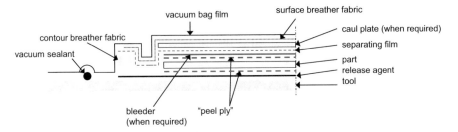

Fig. 4.11 Typical autoclave vacuum setup

normally coated with a release agent in order to avoid unwanted adhesion of the part or its resin. In some cases it can be necessary to use a separating film for this purpose, too.

The surfaces of the CFRP part are usually covered by a perforated separating film, allowing the air or volatiles still included in the lay up to be evacuated during cure. This "peel ply", for example a woven nylon, which must be carefully analysed and qualified for this process, will be removed after cure and prior to downstream processes, and can help to provide clean and activated part surfaces ready for subsequent bonding or coating processes, see Chap. 6.

Depending on the type of prepreg used (with or without excess resin), and depending on the process parameters, it can be necessary to provide a bleeder material (usually a heavy weight polyester fabric), absorbing any excess resin during the process. Another film, for example Nylon® polyamide, separating the lay up from a caul plate, can be placed on top of the peel ply. Caul plates (metal or CFRP), providing a flattened, smooth surface, can be used for certain applications or only at certain locations of a part. Pressure pads, usually silicone parts, can also be applied locally in order to guarantee a sufficient and homogeneous pressure application. This can be necessary especially at edges or in corners; in particular if already cured parts are placed on non-cured prepreg laminates in co-bonding processes.

A surface breather fabric (usually polyester) is applied underneath the vacuum bag film, providing a channel for the air evacuated out of the CFRP laminate to the vacuum connection. For this purpose, the surface breather fabric is connected to the contour breather fabric. Finally, the vacuum bag film, for example a polyamide film, is placed on top of the complete lay-up, and a flexible tape is used to seal the part circumferentially. The vacuum film must provide enough flexibility to ensure an even pressure distribution during compaction. Compression forces applied during the evacuation and during the application of autoclave pressure will lead to laminate compaction, and depending on the part thickness and part size this can result in large displacements (some mm to some cm; the typical thickness tolerance for the cured part thickness of CFRP is 10 %). In case of a non-flexible vacuum film, insufficient pressure would be applied on the CFRP part, or the vacuum film would even tear and lead to waste. The necessary flexibility of the vacuum film is usually achieved by a number of folds. These folds provide excess material and will unfold as the compression proceeds.

Prior to the transfer of the vacuum bagged part into the autoclave, the vacuum tightness is examined carefully. The air underneath the vacuum film is evacuated by means of vacuum pumps, and the resulting pressure is monitored. Figure 4.12 shows a possible curing cycle (as should) for a 180 °C curing epoxy prepreg material. Due to the fact that tools for wing or fuselage skins can be large and thermally inert, the typical heating rates are limited (approx. 0.3 °C/min in Fig. 4.13). With a curing time of 3 h at 180 °C and an additional prior holding phase at a lower temperature the total cycle time in the autoclave can sum up to 8 h or even more.

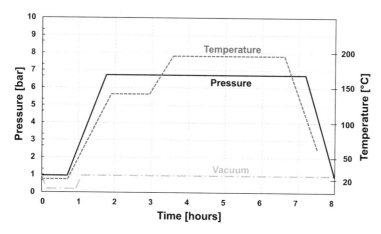

Fig. 4.12 Autoclave curing cycle process parameters (example)

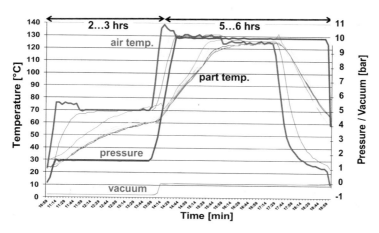

Fig. 4.13 Autoclave process parameter monitoring

The vacuum pump is usually turned off when the pressure is applied in the autoclave in order to protect the pump being damaged by hot gas in case of the failure of the vacuum film. In many cases a pressure of 7 bar or more is applied to guarantee a full compaction of the laminate to the desired thickness, fibre volume content and impregnation quality. Ideally, the laminate should be completely free of voids, although 2 vol% of void content is usually tolerated and proven acceptable for many parts. Figure 4.13 shows the process data recorded during the curing cycle of a 125° curing prepreg part. The part temperature, which must be monitored by thermocouples, varies at different locations of the part, as the heat transfer is inhomogeneous: The hot gas velocity varies within the autoclave, and in addition, the part thickness and the thickness of the vacuum auxiliary materials can vary, too. Also caul plates, locally applied pressure pads or tool thickness changes can lead to inhomogeneous temperature distribution. Due to the thermal inertia, the part temperature follows the air temperature with some delay during heating and cooling.

Once the part is cooled down to an acceptable temperature, the tool can be removed from the autoclave and the part can be debagged. Again, this is a labour and cost intensive manual process, and the auxiliary materials (vacuum film, bleeder, sealant etc.) can only be used once and must be scrapped. New developments deal with alternative reusable vacuum bag material (silicone membranes) for multiple applications but not yet state of the art for large airframe parts [10]. Reusable silicone membranes can be recommended for smaller parts, where handling is not an issue. A cost-performance trade-off taking into account the number of cycles that the reusable bag will last and its initial cost will show the optimum solution.

The CFRP parts are contour milled and quality controlled in downstream processes.

Although alternative process technology for curing is available and has been investigated for decades, the autoclave technology is still dominating in airframe manufacturing. A comparison of process technologies is shown in Fig. 4.14. The domination of autoclave technology is not only due to its universality and flexibility (large autoclaves can be loaded with a large number of different parts of different sizes, even at the same time), but also due to the controllability, which can be very challenging for alternative processes. Also the recertification cost for changes in material and process can be huge and prevent a change of the curing technology.

Composite Defects

In most cases 100 % of the parts manufactured for airframe applications are inspected by non-destructive testing processes. Typical unwanted defects in the cured airframe structure caused by the material or by the manufacturing process are:

• insufficient curing
• porosity

Curing Methods			Material			Applicability				Power (Penetration)			Cost		
		Criteria	Availability	Universality	Potential	Curing/ heating speed	Know-how	Handling-flexibility	Controllability	Polymermatrix	Carbon fiber	Glass/ natural fiber	Investment cost	Running cost	Maintenance cost
Radiation curing		Gamma ray, x-ray	-	+	+/-	+	+/-	--	--	++	++	++	--	--	--
		Electron beam	-	+	+	++	+/-	-	-	+	+	+	-	-	+/-
		Ultraviolet	+/-	+/-	+	++	+/-	++	+	+/-	x	+/-	+	++	++
Thermal curing	Radiation heating	Infrared	++	+	+	+/-	+	++	++	+/-	-	+/-	+	++	++
		Laser	++	+	+	+	+	+	++	+/-	-	+/-	+/-	+	+
		Microwave	++	+/-	+/-	+	+/-	-	+/-	++	-	++	+/-	+	+
	Convection & Conduction heating	Hot gas	++	+	+/-	+/-	+	+	+	-	-	-	+	+/-	+/-
		Flame	++	+/-	+/-	+/-	+	+	+/-	-	-	-	+	+	+
		Oven / Autoclave	++	++	+/-	+/-	+	+	++	-	-	-	+	+	++
		Hot shoe	++	+	+/-	+	+	+/-	+	-	-	-	+/-	+	+
		Induction	++	+/-	+	+	+/-	+	+/-	x	+/-	x	+	++	+
		Ultrasonic	++	+	+	-	+/-	+/-	+/-	++	++	++	+/-	++	+
		Resistance	++	+/-	+	+	+/-	+/-	-	x	+	x	+/-	+/-	+
	Particle-aided heating	Magnetic particles heating	++	+	+	+	+/-	+	+	+	x	+	+	+	++
		NIR particles heating	+	+	+	+	+/-	+	+	+	-	+	+	++	++

(++) very good; (+/-) good; (-) bad; (--) very bad; (x) not applicable

Fig. 4.14 Curing methods comparison [11]

- delamination
- foreign object inclusions
- resin rich areas
- too high fibre volume content
- gaps between rovings or adjacent prepreg layers
- unwanted overlaps
- missing rovings
- fibre misalignment
- wrinkles

It must be secured by appropriate inspection methods during manufacturing that these defects are within the acceptable limits specified during qualification and certification. Otherwise the part must be repaired or scrapped.

Figure 4.15 provides an overview of inspection test methods [12]. Ultrasound inspection is widely used to detect delamination, foreign object inclusions and porosity.

Dielectric analysis (DEA) is often applied to monitor cure. Insufficient cure can occur at "cold spots" of the part, see Fig. 4.13. Sufficient cure must be ensured in order to guarantee the performance of the material (strength, media resistance etc.), see Chap. 3.

Typical dielectric sensors, which can directly be applied on the CFRP laminate lay up to monitor the cure during the autoclave process, are shown in Fig. 4.16.

The process works as follows:

- the sample is placed in contact with two electrodes (the dielectric sensor)
- a voltage (excitation, sinusoidal) is applied

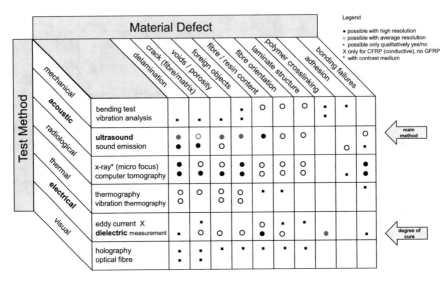

Fig. 4.15 Composite non-destructive inspection methods overview based on [12]

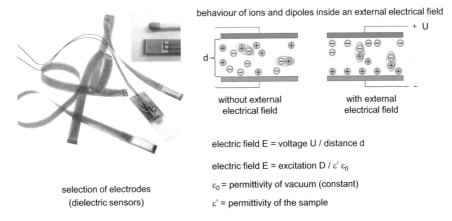

Fig. 4.16 Dielectric sensor and principle [13], photography courtesy of NETZSCH

- the charge carriers inside the sample (the resin) are forced to move
- the movement results in a response current with a phase shift
- the charge carriers are mainly ions, but also dipoles (dipolar groups of the resin)
- the phase shift is a function of the ion mobility and the resin viscosity
- ion mobility decreases with higher viscosity
  - → the response current amplitude is reduced, the phase shift increases
- the amplitude is correlated with the dielectric permittivity ε'
- the dielectric loss factor ε'' is calculated from the phase shift, representing the energy needed to align dipoles and move ions, and reciprocally proportional to ion viscosity ν

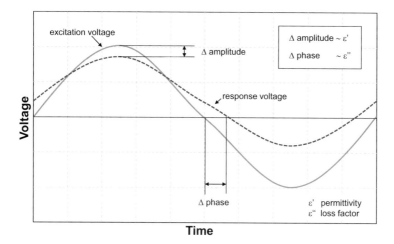

Fig. 4.17 Dielectric measurement voltage excitation and response signal as a function of time

- the loss factor (ion viscosity) is monitored to understand the curing (hardening) process within the autoclave

Figure 4.17 shows the voltage excitation and response signal as a function of time.

Figure 4.18 shows the data recorded (ion viscosity and temperature) during the curing process [13]. A good correlation of dielectric viscosity measurements and rheometric measurements was demonstrated in [14].

In cases where DEA is not applied, traveller specimen (,,in process testing") should be used, which can be investigated by destructive testing after cure (interlaminar shear strength tests ILSS, drum peel tests, DSC etc.).

The ultrasonic test works by the pulse echo or thorough transmission principle shown in Fig. 4.19. The incident sound signal is compared to the transmitted (Through Transmission Ultrasound Technology TTU) or reflected sound (Impulse Echo Technology IE). Foreign objects, delamination or voids will influence the transmitted and reflected signals. However, the interpretation of the transmitted and reflected signals requires special knowhow and careful interpretation, and only trained and qualified personnel are allowed to inspect airframe parts. Reference standards, i.e. CFRP parts prepared with artificial defects and CFRP parts fully free of defects are used for comparison and calibration.

In most cases, water is used as the coupling agent to transfer the sound waves into the composite parts ("squirter" technology, Fig. 4.20). It is also possible to work within a water tank. The sound wave frequency depends on the part thickness; typical frequencies are between 5 and 50 MHz. Lower frequencies can be used for sandwich parts. Coupling of ultrasound waves into CFRP through air is difficult due to a very large signal reflection at the CFRP surface and very weak remaining signals within the CFRP [15–18].

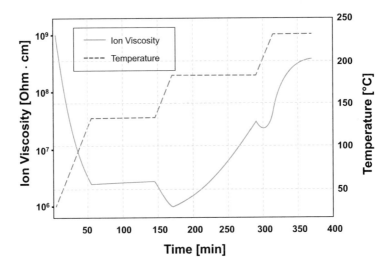

Fig. 4.18 Dielectric measurement data recorded during the curing process [13]

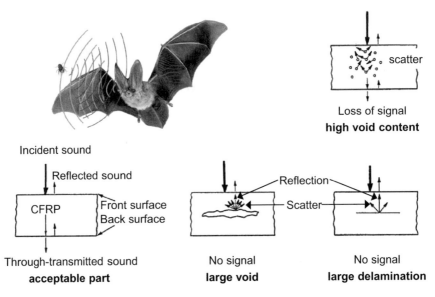

Fig. 4.19 Ultrasonic test and pulse echo principle for delamination and void analysis. Through transmission ultrasound (TTU) compares incident sound and transmitted sound signals. Impulse echo (IE) compares incident sound and reflected sound signals

Large parts are inspected automatically ("squirter" technology). The ultrasonic emitter/receiver scans the complete part surface line by line and the data is recorded ("C-Scan", Fig. 4.21). Information about the position and the total area of the defect is generated. By processing the signals with appropriate software an image can be

Wrong:
No ultrasonic transmission
from emitter to receiver
in case of relative movement

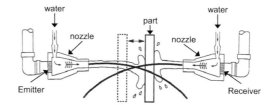

Right:
Ultrasonic transmission
from emitter to receiver

Typical frequencies:
5 to 50 MHz
(thickness dependent)

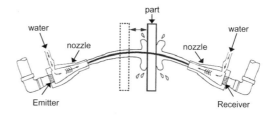

Fig. 4.20 Emitter–receiver principle of ultrasonic sound inspection (TTU) with water coupling ("squirter" technology)

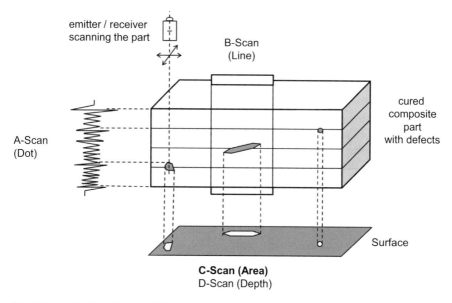

Fig. 4.21 A, B, C and D scans [12]

generated, in which different grades of grey or colours indicate the local signal reduction. A typical result for an impulse echo C-scan (5 MHz phased array probe) is shown in Fig. 4.22. The fibre orientation of 0 °C and 90° is visible, but more importantly, some artificial defects are very clearly visible. It is also possible to determine the depth of a defect, ("D-Scan", Fig. 4.21), i.e. the information about the z-position within the laminate.

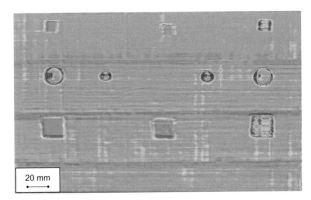

Fig. 4.22 C-Scan image of a 1.45 mm thick CFRP laminate $(0°/90°)_{4s}$ with artificial defects. Delamination squares sized 10 mm × 10 mm (*top*) and 20 mm × 20 mm (*bottom*) in 1.05 mm, 0.7 mm and 0.35 mm depth (from *left* to *right*), drilled holes with a diameter of 18 mm and 10 mm and a depth of 0.3 mm

Special test devices can be necessary for complex shaped structures, especially within radii or in the near field of drilled holes or cut outs. Ultrasonic sensors can also be arranged within arrays, thus providing information for the complete area covered by the array, or focussing on certain areas.

Figure 4.23 summarises important points for the prepreg autoclave technology. High performance and toughened epoxy prepreg materials are available for a wide variety of airframe parts, such as wing and tail plane panels, fuselage skins, stiffeners, frames, cross beams, ribs, brackets etc. However, compared to alternative technologies such as aluminium (see Chap. 3) or the CFRP textile technology (see the following subchapter), material cost are relatively high. Many non-value adding process steps such as storage, transportation, bagging, sealing, debagging and extensive non-destructive testing are today very important contributors to the high manufacturing cost of parts.

New technologies such as reusable vacuum bag material (e.g. silicone membranes) for multiple applications must be further developed. More rapid NDT is needed, and an even better understanding of the maximum allowable defect sizes that can be accepted for a safe design. In addition, although airframe manufacturers can look back on decades of experience with prepreg material and the autoclave technology, further work is necessary to improve the manufacturing process robustness and to work on improved (i.e. a more "forgiving") prepreg material with improved light weight properties such as compression strength after impact in order to reduce quality issues, repair effort and scraped parts, especially for the type of manufacturing defects listed above. Furthermore, thickness tolerances must be reduced. Today's thickness tolerances of about 10 % lead to a high assembly effort, as extra material is needed at part interfaces for compensation ("shim"). Thickness tolerance must be reduced. In addition, autoclave technology simulation

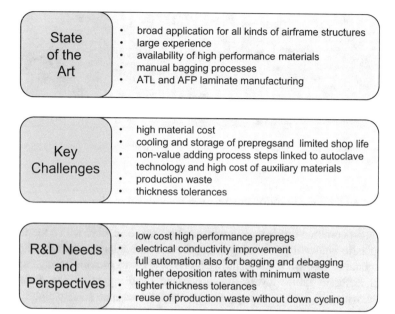

Fig. 4.23 State of the art, key challenges and perspectives of the prepreg autoclave technology

models and software, taking into account several parts cured at the same time, must be further developed to avoid expensive trial and error experiments.

Advanced ATP and AFP machinery is necessary to reduce production waste caused by cuttings and to increase deposition rates, coping with a higher production rate. In addition, new recycling methods for production waste (but also for end of life prepreg airframe parts) and new textile technologies must lead to carbon fibre staple fibre yarn prepregs with a performance that allows a reuse for primary airframe structures. Finally, a prepreg material with improved electrical conductivity could reduce the electrical system installation effort and improve the total weight saving. Possible solutions under investigation are described in Chap. 9.

Textile Infusion Technology

The basic idea of the textile infusion technology is to use the high potential of automated and cost efficient textile manufacturing processes (as developed for decades in the textile and clothing industry) and to combine textile near net-shape preforms with the needed polymer matrix by means of liquid resin infusion. Compared to prepreg, approx. 20 % material cost savings or more are possible, as the impregnation step is not an extra process within the value chain but included in a downstream process at the airframe manufacturer. Different types of carbon (but

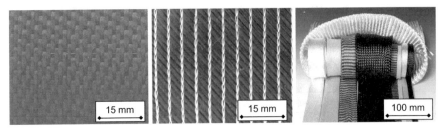

Fig. 4.24 Common textile reinforcements, fabric (*left*), non-crimp fabric (*middle*) and different braided sleeves (*right*)

also glass and aramid) reinforcement have been developed and are commercially available (Fig. 4.24).

The most common one is woven fabric, which can be coated with a thermoset or thermoplastic binder if preforming is required. The binder can be thermally activated to enable preform manufacturing processes, i.e. to provide enough adhesion between different fabric layers for subsequent handling processes. Fabrics are available with different weaving patterns, thicknesses and areal weights. "Non-crimp" fabrics (NCF) have also been successfully developed and qualified for airframe applications.

The difference to woven fabric is the fact that the NCF fibre rovings remain almost straight aligned (where, on the contrary, fibre rovings show typically undulations for warp and weft within a fabric). In addition, multiaxial NCF with several layers in $0°$, $+45°$, $-45°$ and $90°$ direction can be produced, tailored to the design of the component. The individual rovings within a certain layer and the layers themselves are fixed by a sewing yarn.

However, this sewing yarn can lead to local misalignments of rovings, reducing the mechanical performance of the material when compared to unidirectional tape, Fig. 4.25. Also other forms of textiles can be used, such as braided tubes, or even dry rovings (e.g. as filler material in corners). Tailored fibre placement can be used to reinforce highly loaded areas [19]. In general, a high areal weight of the textile can improve deposition rates during manufacturing, however, mechanical laminate properties can be negatively affected.

The most common process to produce airframe parts is the Resin Transfer Moulding (RTM) process. The main steps after the 3-D textile preform has been prepared by prior adequate cutting and binding operations are

- the RTM tool (usually an Invar metal male and female tool) is preheated, typically to 80–120 °C
- dry fibre preforms are positioned within the tool cavity
- the tool is closed, the textile compacted to part size, and the tool is sealed
- the tool is clamped (to prevent unwanted opening during infusion)
- a mould vacuum is applied (typically < 10 mbar)
- the preheated liquid resin is injected under pressure (typically < 7 bar)
- the tool is heated to curing temperature (typically 180 °C)

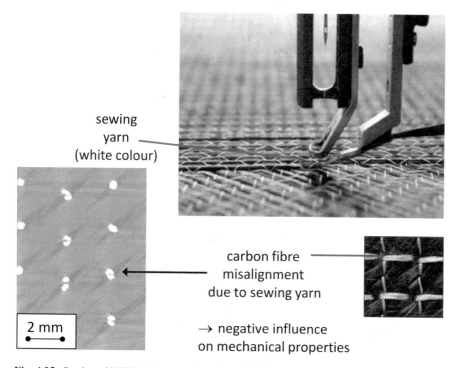

sewing
yarn
(white colour)

carbon fibre
misalignment
due to sewing yarn

→ negative influence
on mechanical properties

2 mm

Fig. 4.25 Sewing of NCF leading to carbon fibre misalignments

- the part is cured (typically for a duration of 90 min, depending on the resin)
- the tool is cooled down (well below T_g)
- the part is demoulded

A common set up is schematically shown in Fig. 4.26. It must be noted that tools and clamping equipment can be very heavy, depending on the pressurised part surface. A pressure difference of 7 bar between the injection pressure within the closed tool and the ambient pressure means a resulting tool opening force of 140 t for a 2 m^2 part surface.

Compared to prepreg autoclave technology, excellent repetitiveness and tight thickness tolerances are achievable (typically 0.1 mm or less) with this isochoric RTM process, and—depending on the tool surface quality—the achievable surface quality is also excellent.

Thick and complex shaped parts with fittings are especially predestined to be manufactured by RTM. Figure 4.27 shows the dry preform and the cured RTM part of a fitting used for the attachment of the vertical tail plane (VTP) to the fuselage. The cured RTM fitting is integrated to the VTP prepreg panels by co-bonding within the autoclave.

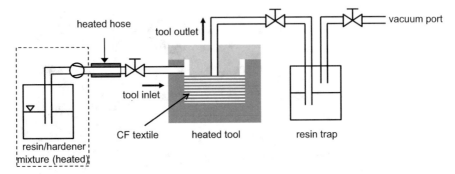

Fig. 4.26 RTM set up

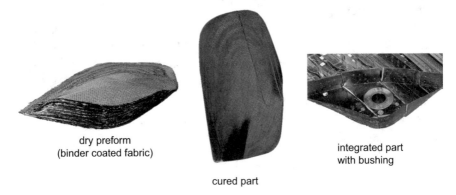

dry preform
(binder coated fabric)

cured part

integrated part
with bushing

Fig. 4.27 RTM VTP attachment fittings [8]

Many variants of the RTM process are existing and qualified, some of them are described in [20, 21]. An excellent description of RTM processes, mould design, component design and troubleshooting of problems can be found in [22].

Important variants commonly applied for series manufacturing of middle sized to large sized (several m^2) airframe parts are Resin Film Infusion (RFI) and Resin Infusion (RI).

For the RFI process, the resin is originally not liquid but provided as a thin film and positioned adjacent to or sandwiched within the reinforcement textile. For the RI process, liquid resin is used, which is poured into the mould in the required quantity, and then the reinforcement textile is placed on top, Fig. 4.28. These processes are particularly suited for thin parts such as ribs. Curing is achieved within the autoclave, or out of autoclave within unpressurised ovens, or by means of heated tools. In the latter case there is only the ambient pressure while the part remains under vacuum, and it has to be analysed case by case if this is fully sufficient to guarantee a void content within the given tolerances.

Thickness tolerances of RFI and RI processes are worse compared to closed mould RTM parts, since with one half of the tool being flexible and not rigid, the

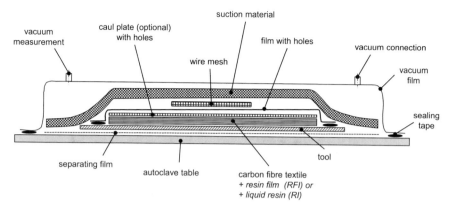

Fig. 4.28 Possible set up for RFI or RI

process is not isochoric but isobar. The fibre volume content is easier to control using RTM.

The resin can be more expensive in case of RFI compared to RTM or RI, as the manufacturing of the film means an additional step in the value chain.

Other advanced resin infusion processes work with semipermeable membranes covering the reinforcement textile, allowing air to be evacuated through the membrane, but holding back the liquid resin (Vacuum Assisted Process VAP®, [23, 24]). VAP-*AP* (VAP with *A*utomatable *P*rocess Setup) is a new automatic process [25–27], VAP-*XF* (e*X*tra *F*ast resin distribution) is a patented new process with a special flow aid, rapidly distributing large amounts of resin on top of the part surface [28, 29].

It is very important for all resin infusion processes to define the process parameters in accordance to the viscosity and curing behaviour of the epoxy resin. Viscosity is a function of temperature and the degree of cure. Depending on the temperature applied, viscosity decreases if the epoxy resin is heated up from room temperature, due to the increasing Braun movement of molecules. For a short period of time (in case of the RTM epoxy resin shown in Fig. 4.29 only a few minutes), it is evident that higher temperatures lead to lower viscosity. A low viscosity is preferable for a fast and complete impregnation of the textile. However, molecule cross linking is also activated and accelerated with higher temperature, causing the viscosity to rise. Depending on the part size a certain time is needed for the liquid resin to fully impregnate the reinforcing textile. Hence, the temperature has to be adequately chosen in order to prevent too much interlacing and a resulting viscosity increase that would no longer allow full impregnation. In many cases it is sufficient to lower the viscosity of the resin below 50 mPas (for comparison: water has a viscosity of approx. 1 mPas at 20 °C, and olive oil 100 mPas at 20 °C). In case of HexFlow® RTM6, an injection and infusion temperature of 120 °C guarantees a time span long enough for most applications.

For the subsequent curing process a high degree of interlacing is desired for optimum material properties, especially under hot/wet conditions. The glass

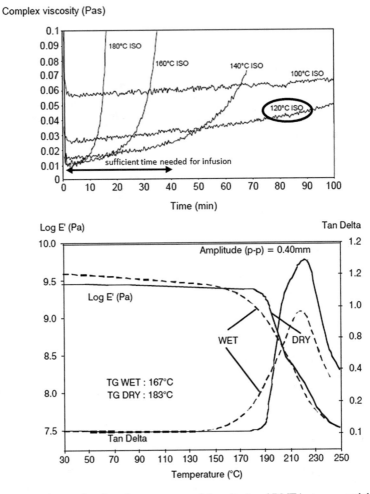

Fig. 4.29 Viscosity as a function of temperature and time (*top*) and DMTA storage modulus as a function of temperature (*bottom*) for Hexcel HexFlow® RTM6 [30]

transition temperature T_g is a good indicator for the degree of interlacing and mechanical material performance under elevated temperature. Figure 4.29 shows the storage modulus measured by DMTA for dry and wet condition for a 180 °C cured epoxy resin, with a resulting T_{gonset} (wet) of 167 °C.

Figure 4.30 shows an overview of resin infusion technology applications of A380. An impressive example of a very large resin infusion part in series production is the rear pressure bulk head of A380, closing the rear fuselage passenger compartment (Fig. 4.31). A similar technology has been chosen for A350.

While the manual effort for positioning the textile preform in the tool and closing the tool is rather limited for the classical closed mould RTM processes, many labour and time intensive steps are necessary in case of single side tool resin

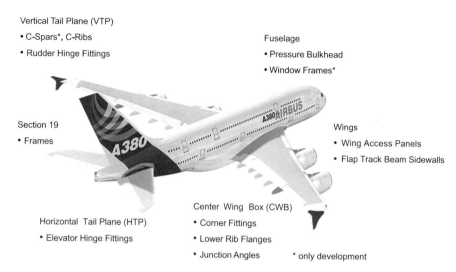

Fig. 4.30 A380 overview of resin infusion technology applications [8]

Fig. 4.31 Rear pressure bulk head of A380 [8]

infusion processes (such as RFI, RI or VAP®). R&D efforts are presently undertaken by the DLR to replace the laser supported manual placement process of large NCF pieces by cooperating robots with end effectors capable of picking, transporting, draping and positioning the textiles [31]. First proposals were however already made in 1986 [32]. In addition, as for the prepreg autoclave technology, R&D is focussing on the full automation of the vacuum bagging [31]. Other new developments deal with the combination of resin infusion and prepreg technology ("SQRTM" = Same Qualified Resin Transfer Moulding"). This process was developed and successfully brought into series production by Radius Engineering Inc. (Salt Lake City, Utah) [33]. SQRTM is a closed moulding method that combines prepreg processing and liquid moulding to produce true net-shape parts in autoclave quality [34].

Fig. 4.32 Wing cover manufactured by the resin infusion technology, European 5th Framework R&D Program TANGO, 2000–2005 [8, 35]

The resin infusion technology has also been thoroughly investigated and tested for wing boxes. Figure 4.32 shows a wing cover developed within the European R&D program TANGO (*T*echnology *A*pplications to *N*ear-term *B*usiness *G*oals and *O*bjectives, from 2000 to 2005). However, it has not been introduced for wing skin series application until today. This is mainly due to the shortfall of material properties compared to prepreg. Due to the higher weight needed for an NCF wing, particularly as a consequence of poor compression strength values compared to prepreg, a business case could not be demonstrated so far even though manufacturing cost advantages were demonstrated. The poor compression strength is caused by fibre undulation, originating in the textile manufacturing process (sewing in case of NCF), and by the relatively low fibre volume fraction.

Figure 4.33 summarises some important facts for the state of the art, key challenges and perspectives of the textile infusion technology. Although material cost savings and manufacturing cost savings are achievable compared to the prepreg autoclave technology, the poor mechanical performance of the material qualified for airframe applications today has limited the spectra of parts. In particular, compared to the latest toughened generation of prepregs, there are shortfalls in compression strength of non-crimp fabrics (approx. 10 % for open hole compression, and 30 % for compression after impact) and bearing strength (approx. 20 %). As the design of many parts is driven by compression and impact load cases, this means additional structural weight in case of NCF application. A promising approach to reduce unwanted fibre undulation and to increase mechanical properties is the Hexcel NC2® technology [36].

Prepreg resins can be toughened by thermoplastic and rubber particles, providing excellent damage tolerance against impact damages and high compression after

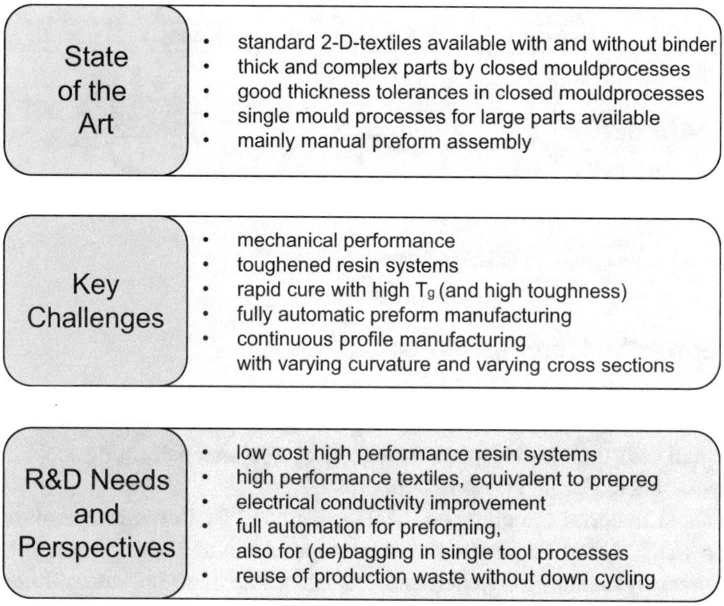

| State of the Art | • standard 2-D-textiles available with and without binder
• thick and complex parts by closed mouldprocesses
• good thickness tolerances in closed mouldprocesses
• single mould processes for large parts available
• mainly manual preform assembly |

| Key Challenges | • mechanical performance
• toughened resin systems
• rapid cure with high T_g (and high toughness)
• fully automatic preform manufacturing
• continuous profile manufacturing
  with varying curvature and varying cross sections |

| R&D Needs and Perspectives | • low cost high performance resin systems
• high performance textiles, equivalent to prepreg
• electrical conductivity improvement
• full automation for preforming,
  also for (de)bagging in single tool processes
• reuse of production waste without down cycling |

Fig. 4.33 State of the art, key challenges and perspectives of the textile infusion technology

impact strength values. However, the toughening possibility of resins suitable for liquid infusion is limited due to the viscosity increase linked to the toughener content and due to filtration effects of thermoplastic particles during the flow process of the resin in the fibre material. R&D is focussing on alternative toughening particles [37]. Also soluble fibres, integrating the textile and working as a toughening agent within the epoxy resin, have been investigated (Cytec Priform TM Technology, [38]).

As production rates will further increase in the near future (see Chap. 1), more rapidly curing resin systems will be needed. These must be chemically designed to be suitable for thick parts (10–100 mm) as well, and it must be insured that their exothermal behaviour can be fully controlled. In addition, they must demonstrate a toughness comparable to prepreg resins. All preforming operations must be fully automatic. Continuous profile manufacturing, also for curved parts with varying radii and cross sections, must be developed. Improved heating and cooling technologies need further investigations. As for the prepreg technology, improved electrical conductivity could help to further improve the light weight potential compared to aluminium structures (integration of electrical functions, see Chap. 9). Finally, similar to the prepreg technology needs, production waste and end of life parts manufactured by the textile infusion technology must be reused for similar purposes.

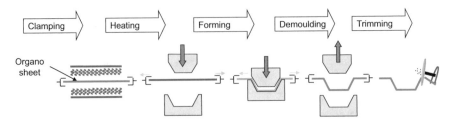

Fig. 4.34 Basic process steps of stamp forming

Thermoplastic Composite Technology

The majority of thermoplastic airframe parts for structural applications are relatively small and thin-walled parts produced by the stamp forming technology in short cycle times. For this purpose, cut outs of
organo sheet material are clamped, heated, softened (with the temperature of the polymer matrix being above the melting temperature) and stamp formed in a fast closing press by means of matched metal moulds. The thermoplastic polymer cools down and consolidates under pressure in the closed and tempered mould. When the press opens, the formed solid part can be demoulded, and the press can be used for the next part. Typical cycle times can be below 2 min per part. Figure 4.34 schematically summarises the main steps [6, 39–41].

Different types of organo sheet material are commercially available on the market and have been qualified for airframe applications. PEEK and PPS are the most common polymers (see Chap. 3), and several layers of different types of fabrics can be impregnated with these polymers in continuous or discontinuous processes into flat sheets of approx. 0.3–5 mm thickness (or more). The double belt process technology and interval hot press technology used to produce these materials is described in [21]. If short cycle times (only a few minutes per part) must be achieved, rapid cooling and re-consolidation is required within the tempered metal mould under pressure (usually several bar), not leaving enough time for inter-fibre matrix flow and elimination of voids within the individual rovings. For a fast stamp forming process of high quality which gives parts free of voids it is thus important to use organo sheet material which has already been entirely micro-impregnated, i.e. each individual filament has been impregnated by the thermoplastic polymer.

In order to shape the organo sheet it is necessary to heat and soften the polymer, in case of PEEK well above 380 °C, and in case of PPS above 290 °C. Different heating devices based on convective, conductive or radiation heat transfer can be used, but infrared radiation applied to both surfaces is most commonly used in production [6]. Infrared heating rates for organo sheet with PEEK or PPS matrix can typically range from approx. 2 to 7 °C/s, depending on the laminate thickness.

The time period between the end of the heating phase and the beginning of the forming process has to be taken into account, as the organo sheet will cool down by convection and radiation immediately after the heating stops. The thermal energy

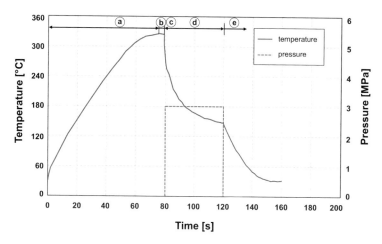

Fig. 4.35 Typical temperature and pressure stamp forming cycle for a 2 mm thick CF/PPS organo sheet, phase (a) heating, phase (b) transportation, phase (c) forming in approx. 1 s, phase (d) - re-consolidation and cooling under pressure in a tempered tool, phase (e) demoulding and cooling to room temperature

stored in the organo sheet must be sufficient to keep the entire laminate above melting temperature for the forming process, allowing all necessary forming mechanisms. This can be achieved by heating the laminate well above the melting temperature range of the PEEK or PPS polymer; however, degradation of the polymer should be avoided.

A typical temperature and pressure cycle is shown in Fig. 4.35. The control of the cooling rate (by adjusting the tool temperature, see phase (d) in Fig. 4.35) is important for the final degree of crystallinity of semi-crystalline polymers such as PEEK and PPS.

By means of automatic transfer devices between the IR-heating station and the press forming station or by means of robots transferring a clamping device containing the heated organo sheet from the IR-heating station to the press forming station, the transfer time can be minimised to only a few seconds.

Figure 4.36 illustrates the key forming mechanisms. The most important forming mechanisms for fabric reinforced thermoplastics are

- shear deformation and
- inter-laminar slip.

Inter-laminar slip is necessary for laminate bending over radii. Inhibiting inter-laminar slip can lead to unwanted distortion, delamination, fibre misalignment, fibre overlapping, local over-pressing, or even to fibre damages. Inter-laminar slip can be inhibited if the forming temperature is too low, if compression forces acting perpendicular to the laminate plane are too high, or if the adjacent layers within the laminate are mechanically fixed.

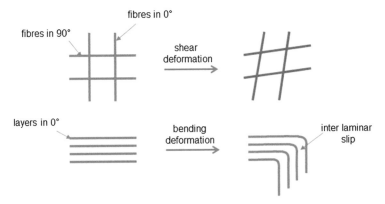

Fig. 4.36 Examples of forming mechanisms of fabric reinforced thermoplastic sheet material (organo sheet); shear (*top*), inter-laminar slip (*bottom*)

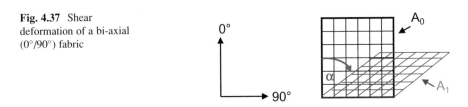

Fig. 4.37 Shear deformation of a bi-axial (0°/90°) fabric

Shear deformation is necessary for draping. While materials with plastic deformation capability such as aluminium sheet can be stretch-formed, locally generating a larger surface by reducing thickness, this is not possible for multi-axial endless fibre reinforced sheet material, since the reinforcing carbon (or glass) fibres are not ductile and their maximum elastic elongation is limited to approx. only 2 % (and less than 5 % for glass fibres) or less.

However, shaping of multi-axial endless fibre reinforced sheet material is possible by shear deformation and thickness increase, Fig. 4.37. The lattice represents fibre bundles in 0° and 90°. The original local area A_0 is diminished to A_1 as a function of the shearing angle α, Eq. (4.1).

$$A_1 = cosa \times A_0 \qquad (4.1)$$

$$V = A \times t \qquad (4.2)$$

$$V = const \qquad (4.3)$$

$$t_1 = \frac{A_0 \times t_0}{A_1} \qquad (4.4)$$

$$t_1 = \frac{t_0}{cosa} \qquad (4.5)$$

where

A_0 is the original fabric area [mm^2]
A_1 is the sheared fabric area [mm^2]
V is the fabric volume [mm^3]
t is the fabric thickness [mm]
t_0 is the original fabric thickness [mm]
t_1 is the thickness of the sheared fabric [mm]
α is the shearing angle [°]

Since the organo sheet can be regarded as an incompressible medium (usually the fibre content is well above 50 vol% and the matrix flow can be almost neglected), the volume of the sheared section remains unchanged, Eqs. (4.2) and (4.3). The thickness t_1 of the sheared organo sheet can therefore be directly calculated by Eq. (4.4). As A_0/A_1 can be set to $1/\cos \alpha$ (Eq. 4.1), the new thickness can be calculated from t_0 as a function of $\cos \alpha$, Eq. (4.5). A 45°-shear deformation of a 2 mm thick organo sheet means a new thickness of 2.8 mm, or an increase of more than 40 %.

If shear deformation, i.e. the rotation of adjacent unidirectional oriented layers relatively to each other, is inhibited during draping, unwanted wrinkling (overlapping of layers), delamination, fibre misalignment, local over-pressing or even fibre damage is likely to occur. Shear deformation can be inhibited, for example if the forming temperature is too low, compression force perpendicular to the laminate plane are too high, or if the adjacent layers within the laminate are mechanically fixed. In case of fabrics, the weaving pattern with the crossings of warp and weft rovings is very important, since the crossings represent a certain "fixture" of the 0° and 90° layer. In order to control shear deformation, it is very helpful to apply membrane tension to the organo sheet during the entire draping process. This can be achieved by adequate clamping and blank holder devices fixed at the outer edges of the organo sheet, providing membrane tension and allowing the drawing-in of the material. Details can be found in [6]. Figure 4.38 shows a

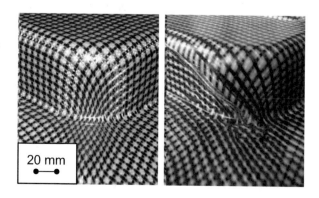

Fig. 4.38 Shear deformation of a fabric-reinforced organo sheet free of wrinkles (*left side*) and with visible wrinkle formation and fibre misalignment (*right side*)

20 mm

complex shaped section of a corner. For a well-controlled process and a sufficient shear deformation the final laminate is free of wrinkles (left side). However, for insufficient shear deformation, unwanted wrinkles and fibre misalignments occur (right side).

Simulation tools, supported by experimental investigations of textiles or heated organo sheet in shear frames ("picture frame test"), can help to identify areas critical for wrinkle formation in complex shaped parts [42]. These shear experiments can be used in order to define the shear locking angle for a given material. This must not be exceeded in order to avoid wrinkling, and is an important input parameter for the simulation and the topological optimisation of the part geometry.

With finite element analysis (FEA) based models fibre reorientation and the resulting shear angles as well as the influence of local stitching on fibre movement can be accurately predicted. One modelling approach is to use a combination of "membrane" and "truss" elements to describe a repeating unit of a reinforcement fabric whose fundamental deformation mechanism is shear.

However, since the ability of the material for a proper re-alignment of its originally (in $0°$, $±45°$ and $90°$ direction) straight aligned unidirectional fibres (in order to fully cover a complex 3-D-shaped surface free of overlaps and wrinkles) depends also on the temperature-dependent viscosity of the thermoplastic matrix, it is necessary to include the temperature dependent material behaviour in the simulation models [43]. Thermal "beam" elements rather than truss elements allow the exact temperature dependent bending behaviour of the material to be characterised and modelled. In addition, an extra layer of thermal "shell" elements allows temperature dependent shear behaviour to be defined. The validation of the temperature dependent shear and bending behaviour requires extensive material testing at non-isothermal conditions.

Although many simulation tools are already commercially available, the temperature dependent modelling of the key forming mechanisms is not yet fully satisfying. For relatively thick laminates (4 mm or more), the cooling rates can be very high near the laminate surface after initial contact with the metal mould ($>100\ °C/s$). This leads to a large viscosity increase of the thermoplastic polymer, inhibiting fibre realignments, while the temperature remains almost constant in the middle of the laminate throughout the complete forming process, which in many cases takes less than a few seconds. The implementation of these non-isothermal effects across the laminate thickness requires extremely long computing times, and even with powerful computing equipment a single draping simulation of a small part can take several days.

For a weight optimised part design, the manufacturing influenced local fibre reorientations and resulting strength and stability properties must be taken into account in subsequent structural simulations, too. In addition, simulation tools must also be used to calculate the local thickness distribution, which is an important input for tool design.

Matched metal moulds (usually made of steel) must provide an accurate cavity which compensates for the final part thickness. Due to the fact that organo sheets with high fibre volume content are almost incompressible, local over-pressings as a

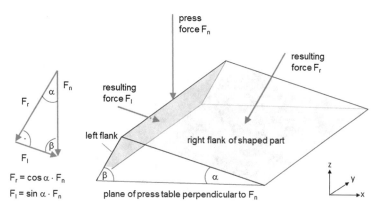

Fig. 4.39 Resulting press forces as a function of part geometry; the pressure distribution can be equalised by adoption of α and β by pivoting of the tool around the y-axis

result of an inadequate tool cavity thickness will inevitably lead to adjacent under-pressings of the laminate, resulting in delamination and poor mechanical properties. In order to achieve a homogeneous pressure distribution, tools with hard stops, i.e. enforcing a defined cavity in the closed state should be avoided. Good results can be expected if the cavity thickness tolerances of matched metal moulds in the closed state for thin (i.e. few mm thickness) parts is below ±0.025 mm. For this purpose, it is important to control the waviness of the tool surface.

Depending on the part geometry, it can be necessary to adjust the position of the tool relatively to the direction of the resulting pressing force by pivoting it around the y-axis, in order to achieve similar pressure distribution on all surfaces of the formed part, Fig. 4.39.

For an even pressure distribution it is also possible to use flexible tools, i.e. to combine a metal mould with a silicone rubber mould, however, cycle times (cooling rates), service life of the tooling (abrasion, embrittlement) and part surface (print through of rovings) will be negatively affected.

Good examples of stamp formed organo sheet are ribs made by Stork Fokker for A340-600 J-Nose, an important part of the wing leading edge, Figs. 4.40 and 4.41 [44]. Stork Fokker moulds these parts from glass fibre reinforced PPS, trademarked Cetex®, available in pre-consolidated organo sheets from TenCate Advanced Composites (Nijverdal, The Netherlands). A similar process was chosen also for A380 J-Nose ribs [45].

Another interesting example is the clip used to connect frames, stringer and fuselage skin for the A350, Figs. 4.42 and 4.43. More than 5000 clips are produced per aircraft. The concept has been developed for carbon fabric reinforced PPS and PEEK organo sheet material from TenCate [46–48].

Organo sheet material and the stamp forming technology is also used for the Dassault Falcon 7X flaps and ailerons, Fig. 4.44.

Fig. 4.40 Press forming of stiffened ribs for A340-600 J-Nose [44] (photography courtesy of Fokker Aerostructures)

Fig. 4.41 Assembly of press-formed stiffened ribs for A340-600 J-Nose [44] (photography courtesy of Fokker Aerostructures)

An excellent overview of organo sheet applications for primary airframe structure can be found in [50]. The development of thermoplastic control surfaces is described in [51]. Also floor structures have been developed and applied [52, 53].

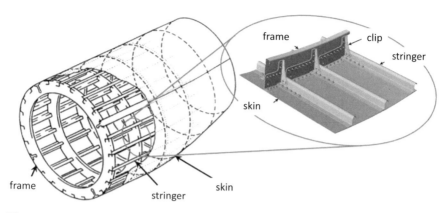

Fig. 4.42 Monolithic fuselage design scheme example with clips in order to connect fuselage skin, fuselage frames and stringer profiles

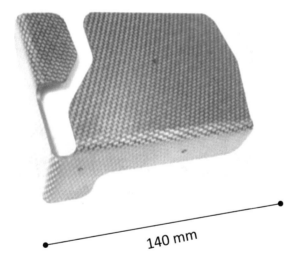

Fig. 4.43 Stamp-formed clip made out of carbon fabric reinforced PPS [photography courtesy of Premium Aerotec]

As the size of parts that can be made by the stamp forming technology is limited to the size of the available presses and tooling, efforts have been made to develop the tape laying technology with thermoplastic carbon fibre tapes [54, 55].

The basic principle is similar to the tape laying technology described earlier in this chapter: A pre-impregnated tape is placed on a tool and cut, and then adjacent layers are placed until the surface is fully covered and the next layer of the laminate can be placed with the desired fibre orientation, Fig. 4.45. A key difference to the thermoset tape is the fact that the thermoplastic material has to be heated above the

Fig. 4.44 Dassault Falcon 7X flaps and ailerons with fabric reinforced thermoplastic manufactured by Fokker [49]

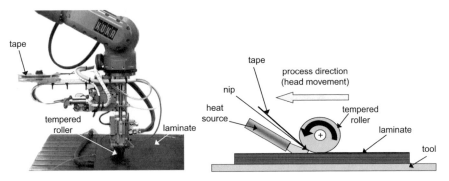

Fig. 4.45 IVW thermoplastic tape laying head "Evo I" mounted on a 6-axis robot (*left*), tape placement principle (*right*) [21]

melting temperature before its placement on the tool surface or on top of the layer of tape underneath. Different heat sources are possible (infrared radiation, hot gas, hydrogen flame, laser, see [21]). Cooling the material again under pressure by means of a tempered metal roller then allows welding the laminate layer by layer. On lab scale, the technology has already demonstrated its capability to produce high quality parts free of voids with carbon fibre reinforced PEEK, without subsequent autoclave cure. However, the achievable deposition rates are not yet satisfying for an economic process, although process speeds of 6 m/min with laser (1.6 kg/h for a single tape with only 12 mm tape width and a tape thickness 240 μm) and 3 m/min with hot gas could successfully be demonstrated on lab scale. Especially for complex shapes, such as wing skins or rear fuselage skins, it can be difficult to control defined cooling rates by means of a hard tempered metal roller at high processing speed (high deposition rates), since an even and linear contact of the roller surface in the tape welding zone can convert into single contact points for a

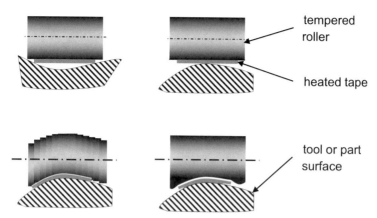

tempered roller

heated tape

tool or part surface

Fig. 4.46 Roller–tape–interface. Hard roller on a concave tool surface (*top left*), hard roller on convex tool surface (*top right*), segmented roller on a convex tool surface (*bottom left*), roller with flexible coating on convex tool surface (*bottom right*)

convex or concave tool or part surface, with an inadequate pressure distribution and uncontrolled temperature gradients, Fig. 4.46. This can lead to delamination and high void content of the part. Using very small tapes can mitigate this problem but will decrease deposition rates. Ongoing research work with segmented rollers or flexible roller coatings has to be continued in order to investigate the best solutions for future series applications.

The ability of thin tapes (with a thickness of typically 250 μm or less) with high fibre volume content (typically 60 %) to deform under the pressure of the roller after being heated above the polymer melting temperature is very limited. Thus, increasing tape width by reducing tape thickness is only possible within very small borders. It is therefore necessary to control the accuracy of the tape placement in x/y-direction in order to avoid unwanted gaps (or overlaps) between adjacent tapes, lowering material performance. Tight tolerances of the width of the tapes used for the placement process can help to mitigate this problem.

Best results at highest processing speeds resulting in high quality laminates free of voids can be expected if the original tape material is already fully micro-impregnated.

Figure 4.47 summarises some important facts for the state of the art, key challenges and perspectives of the thermoplastic technology. Qualified carbon fibre reinforced PEEK and PPS organo sheet material is commercially available for stamp forming processes. However, compared to epoxy prepregs, material cost are relatively high, and the size of the semi-finished material as well as the size of the parts manufactured by the stamp forming technology is limited.

A promising new development presently under research is the combination of stamp forming and compression moulding or injection moulding: Multiaxial endless fibre reinforced areas of a given part are combined with discontinuously reinforced areas [56]. As discontinuously fibre reinforced thermoplastics are able

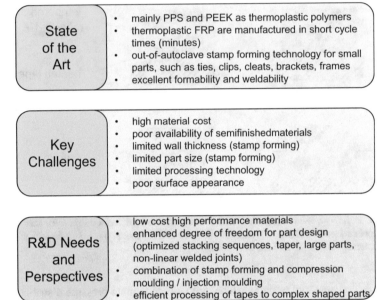

Fig. 4.47 State of the art, key challenges and perspectives of the thermoplastic technology

to flow when heated, very complex shaped part sections can be formed, adding part stiffness and functionality (e.g. formation of very thin ribs, smooth surface finish, local load introduction areas, incorporation of metal connection parts etc.). In future, discontinuously fibre reinforced thermoplastics could be made out of production waste or by plasticising thermoplastic end of life parts. Technology to control fibre length distributions of the recycled materials as well as the technology to guarantee reproducible fibre orientation within the relevant sections of the finished parts is mandatory.

Further improvement of the stamp forming technology with organo sheet material could be possible by using impregnated non crimp fabrics, taking advantage of improved mechanical properties (especially compression strength) and higher light weight potential when compared to conventional fabric, and by using organo sheet with a laminate stacking sequence, containing not only 0°/90° oriented fibres, but also ±45° fibres, tailored for the relevant part.

With only a few exceptions, today mainly small parts are produced for series applications, such as clips or brackets. In order to widen the spectrum of potential applications, especially to enable an efficient production of large and complex shaped shells for tail planes, fuselage panels and wings, with optimised stacking sequence, local thickness variations, cut outs, and with an acceptable surface appearance, capturing the benefit of completely avoiding autoclave curing and non-value operations linked to the autoclave, the thermoplastic tape laying technology must be further developed.

Finally, fully automated welding technologies to join large and complex shaped parts need further development. Induction welding is a promising new approach [57], potentially enabling longitudinal and circumferential skin joints as well as skin to stiffener joints.

Filament Winding Technology

Although huge advancements have been made for thermoplastic winding [58], the main technology used to produce rotation-symmetric parts for series application is based on thermoset resin. Examples for rotation-symmetric parts are compression struts supporting the passenger floor beams, tension/compression rods for the attachment of overhead stowage compartments in the fuselage, pressure vessels (e.g. for the hydraulic system) or storage vessels (e.g. for potable water or waste water containment) (Fig. 4.48).

Filament winding starts with the roving which is pulled through a resin reservoir for impregnation and wound on a mandrel (or a "liner") under tension. Alternatively, prepregs (normally with unidirectional fibre orientation) can be used.

The stacking sequence of the desired laminate can be fully controlled by the winding machine, as the position of the roving placement can move along the axis of the rotating mandrel. With a winding speed of up to 100 m/min (in case of prepreg winding) a deposition rate of more than 100 kg/h can be achieved. In a subsequent step, heat must be applied to cure the epoxy matrix. This is usually achieved in convection ovens.

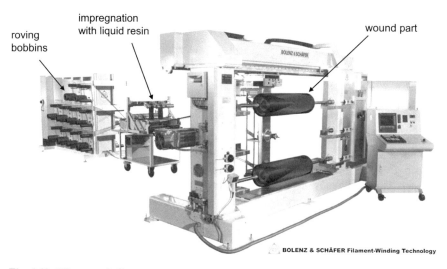

Fig. 4.48 Filament winding

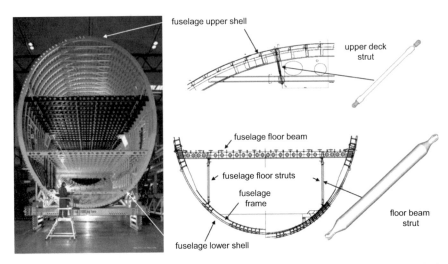

Fig. 4.49 A380 fuselage cross section with passenger floors, floor struts (*bottom*) and upper deck struts (*top*)

Fig. 4.50 A380 CFRP tension-compression struts (photography courtesy of CirComp)

Figure 4.49 illustrates the A380 fuselage cross section. This aircraft has two passenger floors. Both passenger floors are supported by cross beams (horizontal beams). The lower passenger floor cross beams are vertically supported by struts, which are attached to the fuselage frames. Upper deck struts are applied underneath the fuselage crown.

Figure 4.50 shows some examples for filament wound CRFP struts (the left and right one coated with white paint) with metal load introduction parts.

Fig. 4.51 A320 Type II hydraulic pressure vessel 1l/200 bar (photography courtesy of CirComp)

Figure 4.51 shows a pressure vessel. The metal vessel is reinforced by aramide fibres (note that the circumferential orientation is clearly visible, supporting the circumferential stresses according to the boiler formula).

Figure 4.53 summarises some important facts for the state of the art, key challenges and perspectives of the filament winding technology.

Recent R&D has focused on the improvement of the deposition rates [59]. A new ring winding head has been developed, capable of processing 12×4 24k CF-rovings simultaneously, Fig. 4.52. A very fast micro-impregnation is achieved by a siphon-impregnation unit, which is constantly fed with liquid resin. Each winding arm is attached to one impregnation unit, and each unit can accommodate four rovings. High winding speeds of up to 30 m/min could successfully be demonstrated (approx. 4 kg/h), resulting in laminates with a void content of less than 2 vol%.

R&D is also carried out to investigate advanced concepts for load introduction. A special challenge is to improve the energy absorption behaviour of floor struts under compression, eventually supporting the crashworthiness behaviour of the fuselage structure [60]. Different devices have been investigated to trigger highly energy absorbing destruction mechanisms of the CFRP.

Latest R&D deals with the development of low-cost high-performance resin systems [61]. The main components consist of high temperature resistant cyanate ester, epoxy resin, catalysts and nano-scaled toughening agents (approx. 20 mass%) Although the toughener content is high, the new resin formulations are free of solvents and can be processed at moderate temperatures of 50–100 °C at low viscosity (<1 Pa·s). The glass transition temperature (T_g) can reach up to 420 °C, leading to very high operating temperatures of up to 400 °C. This development can enable new CFRP applications in temperature loaded areas.

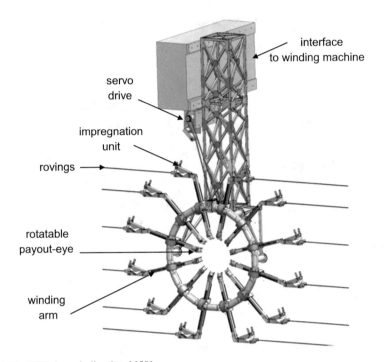

Fig. 4.52 IVW ring winding head [59]

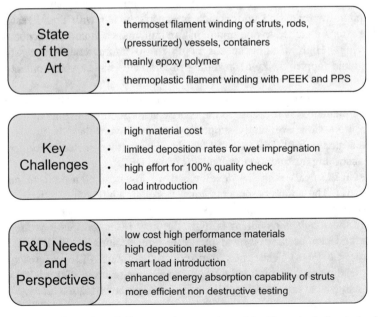

Fig. 4.53 State of the art, key challenges and perspectives of the filament winding technology

Finally, as for all other manufacturing technologies, the non-destructive test effort for a 100 % test of all parts (mainly for void content and delamination) has a significant influence on manufacturing cost, and advanced test technology is needed to accelerate and ease testing.

Joining and Bonding

Bolted Joints

The dominating technology to integrate a number of single CFRP parts to complex subassemblies, sub components and major components is bolting. Titanium fasteners are most common due to their excellent corrosion resistance. The list of advantages of this joining technology is long:

- large experience
- high world wide availability of technology and expertise (also at repair shops)
- very fast processing, fully automatic, very short cycle times
- large variety of qualified rivets and bolts
- relatively simple QA
- dismantling of parts possible
- qualified bolted structural repair methods
- no need of special equipment for temperature or pressure application (as for adhesive bonding)

Figure 4.54 shows a selection of fasteners. These are available in various diameters and lengths, and with different heads. The application of protruded heads is very common for CFRP joints in non-visible areas. Counter sunk heads are used for flush surfaces (protruded heads can produce parasitic drag on surfaces).

Figure 4.55 shows the Hi-Lok® system. "The HI-LOK™ fastening system consists of two parts, a threaded pin and a threaded collar. The collar was designed to consistently repeat preload on each fastener. The wrenching element breaks away at the designed preload. It eliminates the need for torque inspection and torque wrench calibration [62]."

Blind bolts can be used for assemblies from one side only. The assembly principle can be seen in Fig. 4.56. However, the advantage of a one side assembly must carefully be traded against the shortfalls of blind bolts, because they are more expensive, and the open hole left within the bolt head can cause negative optical appearance and negative aerodynamic impact (parasitic drag).

The main load transfer principle for bolted CFRP joints is based on shear. More details are given in Chap. 6 (bolted repair). The principle of a screwed joint, which is to generate normal forces within the screw bolt in order to press part A and B together as to ensure sufficient resistance against a relative slip movement of the parts is usually not applied for CFRP. The reason is that polymers—even if

Fig. 4.54 Selection of fasteners

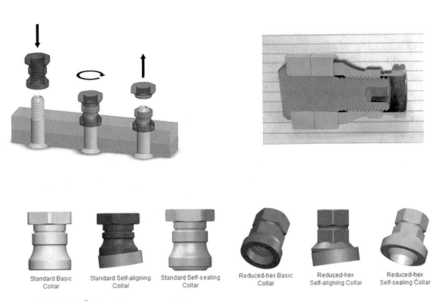

| Standard Basic Collar | Standard Self-aligning Collar | Standard Self-sealing Collar | Reduced-hex Basic Collar | Reduced-hex Self-aligning Collar | Reduced-hex Self-sealing Collar |

Fig. 4.55 Hi-Lok® fastening system [62]

reinforced with high volume fractions of carbon fibres—tend to creep. This would lead to the loss of normal forces within a screwed CFRP joint.

An interference fit is most commonly used when titanium bolts are brought into CFRP. A clearance fit could lead to bolt tilting when the parts are loaded and trying to slip. This would cause high bearing loads of the tilted bolt on the laminate edges.

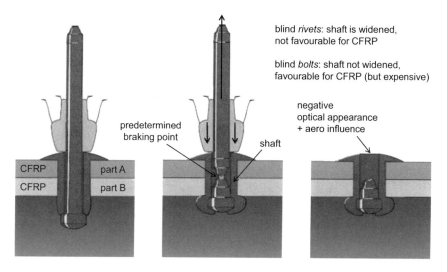

Fig. 4.56 Blind bolt assembly principle, image source (without text) see [63]

A press fit can lead to problems of bolt insertion, as high z-forces on the laminates can lead to unwanted delamination adjacent to the hole.

A possible procedure for a fully automated riveting process is:

- the CFRP parts are coated with sealing on the surfaces of the junction area and clamped
- drilling of holes (e.g. orbital or oscillating drilling with poly-crystalline diamond coated tools)
- while drilling, the chippings are extracted by suction
- if CFRP and metal (aluminium parts) are joined, the metal part should to be on the top side during the drilling operation in order to avoid damage of the CFRP hole by metal chips
- if necessary, countersunk holes can be generated during the same drilling operation
- sealant can be applied from a cartridge to cover the hole surface
- the rivet is inserted
- the head is closed
- the position is unclamped

Today riveting machines can assemble between 15 and 20 rivets per minute, including all above-mentioned steps.

Figure 4.57 shows a fuselage side shell mounted in a Broetje-Automation riveting machine. Figure 4.58 shows a nose fuselage section riveting operation (Broetje-Automation) [64].

A modern aircraft contains more than 100,000 of titanium rivets (an A380 VTP alone contains approximately 10,000 rivets), contributing significantly to the over-all material cost. Although the advantages for bolted CFRP joints listed above have

Fig. 4.57 Fuselage side shell mounted in a Broetje-Automation riveting machine (photography courtesy of Broetje-Automation) [64]

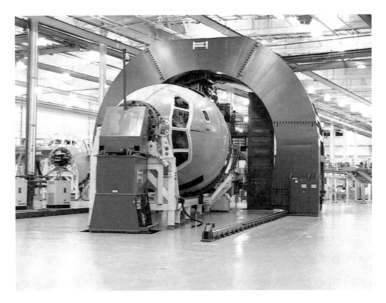

Fig. 4.58 Nose fuselage section riveting operation with 360° fuselage section riveting and automatic collar installation, also for CFRP-metal joints (photography courtesy of Broetje-Automation) [64]

led to the fact that titanium rivets are the number one joining technology, several disadvantages must be highlighted:

- titanium bolts contribute to part weight significantly, depending on the specific part approx. 5 %
- additional weight is caused by incorporating the notch factors, i.e. additional material thickness is necessary due to the holes (see Chap. 3 tables comparing the strength values of notched and plain CFRP laminates); additional material thickness is necessary in all part areas that might later be prone to structural bolted repair
- rivet head pull through of countersunk rivets is an important criteria for the minimum CFRP skin thickness which is achievable today (i.e. other strength and stability criteria could be fulfilled with lower CFRP skin thickness, hence with less weight)
- special lightning strike precautions can be necessary in order to avoid edge glow or sparkling effects close to the bolts, especially in wing boxes and fuel tank areas
- titanium rivets are expensive; depending on the specific part the rivets alone can make a portion of approx. 5 % of the total part cost

Adhesive Joining

The integration of CFRP parts by the adhesive function of the (epoxy) resin is a very common process if one or all of the parts to be joined are not yet cured (see "skin/stringer integration at the beginning of this chapter; more details for co-curing and co-bonding operations are explained in Chap. 8). This subchapter is focussing on secondary bonding operations, assuming that the individual CFRP parts are fully cured prior to the joining process.

The most applied adhesive material is based on epoxy films, Fig. 4.59, delivered on rolls in different widths and thicknesses. Different adhesive film materials are qualified for secondary bonding operations, and usually 180 °C autoclave curing is required. The autoclave process technology has been described earlier in this chapter. The films come with a removable backing material and a carrier textile (optional). This guarantees a constant adhesive thickness.

One of the most important steps of the adhesive bonding procedure is the surface pre-treatment [65, 66]. Different ways of pre-treatment are for example

- peel ply removal (state of the art, simple, fast; but: peel ply must be free of contamination and release agents)
- blasting (but: difficult control of material removal rate, dust problems)
- plasma (but: limited material removal rate)
- laser (but: more expensive than peel ply removal)
- grinding (but: difficult automation, and difficult control of material removal rate)

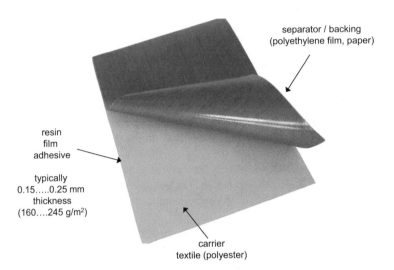

Fig. 4.59 Adhesive film

A qualified and most often applied procedure is the removal of peel ply directly prior to the application of the adhesive (since activated surfaces tend to contaminate when exposed to normal conditions, the maximum time span in between these operations must not be exceeded as matter of the bonding process specification).

Laser stripping is more and more used for surface pre-treatment prior to painting, and large surfaces can be treated in relatively short periods of time (a pulsed 1 kW CO_2-Laser takes approx. 20 min to scan a VTP CFRP outer shell surface).

Grinding is difficult to control. Figure 4.60 shows images of CFRP surfaces gained by scanning electron microscopy at Wehrwissenschaftliches Institut für Werk- und Betriebsstoffe [67]. The ground surface remains „contaminated" with abrasive material even after an air blast treatment, and extensive ultrasonic cleaning is necessary for a complete removal of particles.

However, peel ply can also lead to unsatisfactory bonding results. Peel plies to be qualified for subsequent bonding operations (after their removal from the CFRP parts) must be carefully analysed to ensure that unwanted constituents such as carbon-fluorine-combinations will not remain on the CFRP surface and weaken the structural performance of the bond.

A simple yet powerful test for the readiness of an activated CFRP surface ready for adhesive bonding is the "water break test". The pre-treated CFRP surface is overflown with water. Contaminated areas of the part can be detected where the water film is "broken", Fig. 4.61. The process has been automated for large parts, and visual inspection is performed by qualified operators. However, the parts must be dried prior to the application of the adhesive.

In addition to the surface preparation effort and the accurate monitoring of the curing process parameters, traveller specimen are often prepared (i.e. small specimen with the same material, cured in the same lot under the same conditions as the

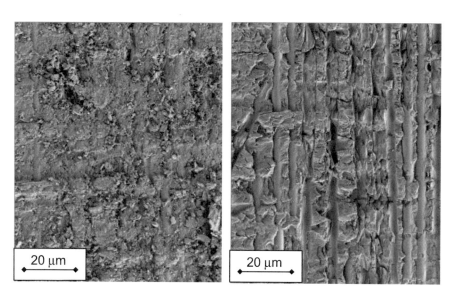

Fig. 4.60 Ground CFRP surface after air blast cleaning (*left*) and ultrasonic cleaning (*right*) [67]

Fig. 4.61 Water break test
principle [68]

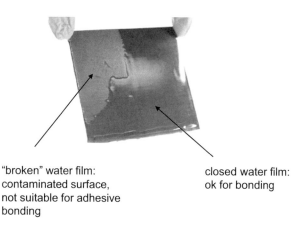

"broken" water film:
contaminated surface,
not suitable for adhesive
bonding

closed water film:
ok for bonding

relevant parts) in order to check the quality of the adhesive joint by means of structural tests, e.g. lap shear tests and Gic-tests. All adhesive joints within a CFRP part are also carefully checked by ultrasonic test. In summary, it must be pointed out that the quality assurance effort for secondary adhesive bonding operations in production is very high. However, due to incidents with unexpected failures of adhesive joints in the past, special design features can be necessary and required by the authorities for certification in order to demonstrate a sufficient load carrying capability of the relevant structure even in case of failure of a certain adhesive bond.

Adhesive bonding is not state of the art for structural repair, see Chap. 6.

New developments deal with full automation of adhesive bonding operations, the application of fast curing and high performance adhesive paste systems, and with the development of detachable bonded joints [69].

Welding

Welding of composite parts is today only applied for thermoplastic CFRP. An overview of the different welding processes can be found in [21]. Welding of CFRP is not widely introduced in series manufacturing of airframe parts, because thermoplastic parts make only a small portion (few mass%) of today's CFRP airframe applications.

Ultrasonic welding uses mechanical oscillations in the ultrasonic region (>16 kHz), leading to polymer melting by absorption and reflection of oscillations as well as by friction within the welding interface [21]. In practise, it is used to fix thermoplastic semi-preg plies prior to autoclave processing [45].

The processing technology of vibration welding of reinforced thermoplastics has been available for decades and very well understood [21]; however, part complexity and component dimensions are very limited, excluding this technology for large airframe parts (e.g. skin/stringer connections, skin joints etc.).

A promising new approach under R&D is the induction welding technology for large and complex shaped parts [57]. Heat is generated by means of an induction coil. The coil is electrically connected to a power station, generating an alternating magnetic field by means of alternating current (typically with a frequency of 400–900 Hz, and a voltage of 150 V). The alternating magnetic field can be used to generate eddy currents, either within a special susceptor material which is placed in the welding zone (a metallic mesh), or within the CFRP laminate itself. Both, eddy currents and the hysteresis of alternating magnetic dipoles, generate heat losses, leading to polymer melt. The joint can be generated by applying pressure to the welding zone and consolidation by cooling. In practise, this can be achieved by a tempered metal roller. Welding heads, containing the induction device as well as the tempered rollers, can be mounted on robots in order to weld complex curvatures continuously or by spot welding. The technology can also be used during the pre-assembly of parts, providing sufficient fixture and strength for handling operations prior to bolted assembly operations.

The present research work is focussing on the temperature control in the welding zone and the maximisation of the welding speed. For continuous welding, speeds up to 0.5 m/min have successfully been demonstrated for 2 mm thick CFRP parts in the lab.

Presently ongoing R&D on thermoset welding with parts made by the autoclave-prepreg technology, but with thermoplastic surfaces, is described in [70].

A quantitative comparison of joining technologies can be seen in Table 4.2. A key advantage of adhesive bonding of CFRP structures compared to bolted joints is the relatively low impact on part weight. However, this potential could only be fully

Table 4.2 Qualitative comparison of joining technologies for primary CFRP airframe structures

Criteria	Bolted joints	Adhesive joints	Welded joints
Experience	++	+	–
Degree of automation	++	–[a]	+
Cycle time	++	–[a]	+
Quality assurance effort	+	–	0
Dismantling possibility	+	–[a]	0
Structural repair possibility	++	–	0
Optical appearance	–	+	+
Impact on drag	–	+	+
Impact on weight	–	+	+
Impact on non-recurring cost	–	–[a]	0
Impact on recurring cost	–	0	+

[a]Adhesive joining refers to materials and processes with adhesive film and autoclave cure; other adhesives and processes can demonstrate improved properties

realised if bolted structural repair of in-service damages could be replaced by adhesive structural repair, which is not state of the art today (see Chap. 6), because otherwise additional material thickness has to be provided anyway within the CFRP structure in order to cope with the notch factors.

The main challenge of adhesive joints for load carrying CFRP airframe structures is the fact that a simple, fast and cost efficient way of ensuring the mechanical performance of the bond for quality assessments in production is not available today.

The negative impact of autoclave cure on non-recurring cost can be mitigated by advanced adhesives with rapid curing capability, and recurring cost can be further improved by a higher degree of automation. The welding technology will further emerge as more thermoplastic matrix materials are used in the future. It has a high future potential for short cycle times, moderate cost for equipment and low recurring cost.

Manufacturing Technology Selection

Figure 4.62 provides an overview of the different manufacturing technologies, classifying possible part sizes ("small" refers to a cm to dm range, "medium" from dm to several m, and "large" means parts with lengths of more than 10 m) and part complexity (where "low" means plane or simply curved structures, "medium" means more complex 3-D-topology with non-developable surfaces, and "high" means integrated stiffened structures). Some typical examples of airframe structures are given for each manufacturing technology.

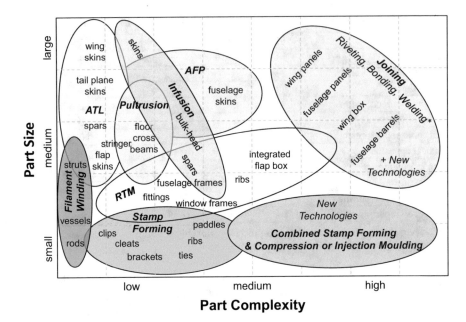

Fig. 4.62 Main manufacturing technologies and airframe applications. Note that "Infusion" also covers open mould and out of autoclave processes. *Continuous CFRP welding is under development

During the pre-development phase it can be helpful to elaborate important criteria in order to find the right manufacturing technology for a given part. For this purpose all important criteria should be identified, listed and prioritised, such as for example (list not exhaustive):

Related to part properties

- part size
- part thickness
- topological complexity (flat, single curved, double curved, undercuts,…)
- tolerances
- lightweight design optimisation (fibre straightness, stacking sequences, fibre volume content, ramp rates…)
- surface smoothness
- …

Related to part quantity and cost

- qualification effort
- automation potential
- cycle time
- energy consumption
- fly to buy ratio

- effort of further treatment (milling, ...)
- quality assurance effort
- tool cost (non recurring cost)
- ...

Related to raw material, semifinished material and auxiliary material

- material cost
- material delivery form (width, thickness...)
- material availability (single source, dual source, ...)
- auxiliary material cost
- ...

Related to resources

- availability of existing production facilities
- cost of new facilities
- availability of skilled staff
- availability of suppliers (make instead of buy)
- ...

Other criteria

- flexibility/compatibility of the manufacturing process in case of late design changes of the part
- part handling
- part recycling
- life cycle analysis, CO_2 footprint, grey energy, ...
- ...

A possible way to rank and assess criteria for a given part is shown in Table 4.3. Each criterion can be linked to a weighting factor f, a value x_{PC} for the part requirement and a value x_{MC} for the manufacturing technology property. If, for example, a certain part requires a very tight thickness tolerance, the value x_{PC} could be set to "3". Due to the fact that RTM delivers tight thickness tolerances, the value x_{MC} for this manufacturing technology could be set to "3". Since compared to RTM the thickness tolerances delivered by ATP is usually worse, a lower value should be assessed for this manufacturing technology. A comparison of the different manufacturing technologies is possible by defining a level of conformance L_{Cn} for each criterion, building the quotient of what can be expected from the manufacturing technology and what is required by the part, i.e. x_{MCn}/x_{PCn}, Eq. (4.6). By definition, the maximum level of conformance should be set to "1", i.e. all values greater than "1" are reduced to "1". An overall level of conformance L_{CO} for each manufacturing technology can be calculated according to Eq. (4.7).

Table 4.3 Quantitative assessment and comparison of manufacturing technologies

Criteria	Weighting factor $f_n = 1\ldots5$	Part requirement 1 = low, small; 2 = medium; 3 = high, large	Manufacturing technology property 0 = impossible; 1 = low, small; 2 = medium; 3 = high, large	Manufacturing technology					
				M_1 ATP	M_2 AFP	M_3 RTM	M_4 MVI	M_5 Pultrusion	M_6 …
C_1	f_1	X_{PC1}	X_{MC1}						
			Conformance X_{MC1}/X_{PC1}						
C_2	f_2	X_{PC2}	X_{MC2}						
			Conformance X_{MC2}/X_{PC2}						
C_3	f_3	X_{PC3}	X_{MC3}						
			Conformance X_{MC3}/X_{PC3}						
C_4	f_4	X_{PC4}	X_{MC4}						
			Conformance X_{MC4}/X_{PC4}						
C_5	f_5	X_{PC5}	X_{MC5}						
			Conformance X_{MC5}/X_{PC5}						
C_6	f_6	X_{PC6}	X_{MC6}						
			Conformance X_{MC6}/X_{PC6}						
C_7	f_7	X_{PC7}	X_{MC7}						
			Conformance X_{MC7}/X_{PC7}						
C_8	f_8	X_{PC8}	X_{MC8}						
			Conformance X_{MC8}/X_{PC8}						
…	…	…	…						
C_i	f_i	X_{PCi}	X_{MCi}						
			Conformance X_{MCi}/X_{PCi}						

Resulting overall conformance $L_{CO} = \sum(f_n X_{MCn}/X_{PCn})/\sum f_n$

$$L_{Cn} = \frac{x_{MCn}}{x_{PCn}} \leq 1 \tag{4.6}$$

$$L_{CO} = \frac{\sum\limits_{n=1}^{i} \frac{f_n \times x_{MCn}}{x_{PCn}}}{\sum\limits_{n=1}^{i} f_n} \tag{4.7}$$

where

L_{Cn} level of conformance for a certain criterion n (0. . ..1)
x_{MC} manufacturing technology property assessment for a certain criterion
x_{PC} part requirement for a certain criterion
f weighting factor f of a certain criterion
L_{CO} overall level of conformance for a certain manufacturing technology and a certain part (0. . ..1)
n number of criteria

This systematic assessment of criteria can also be important to identify no-go criteria, for example, when semifinished material cannot be made available in the required size. In this case the value for x_{MC} should be set to "0".

However, the method described here can only be useful in an early stage of the pre-development, and full cost analysis of non-recurring cost and recurring cost as well as weight assessments are necessary for final trades of preferred manufacturing technologies and the selection of an optimised route.

Questions

1. Why is prepreg usually stored at $-18\ ^\circ$C and preferably processed in air-conditioned rooms?
2. What is the difference between ATL (Automated Tape Laying) and AFP (Automated Fibre Placement)?
3. How are prepreg components typically prepared for curing in an autoclave?
4. What is „co-curing", „co-bonding", „secondary bonding"?
5. Why the vacuum is usually turned off in the autoclave when overpressure builds up?
6. Which continuous manufacturing process can be used for continuous profiles such as stiffeners?
7. What is a typical temperature/pressure profile of the autoclave for the most commonly used prepregs?
8. What contributes significantly to the limitation of the achievable heating rates of the components to be cured in an autoclave?

9. What are some typical composite defects that may occur based on material or production conditions?
10. Which method is commonly used to monitor the curing of the epoxy resin of the components in the autoclave?
11. How does dielectric analysis work?
12. Which method is typically used for 100 % inspection of CFRP components concerning porosity, delamination and foreign inclusions etc.?
13. How does ultrasonic testing of CFRP components work?
14. What is a so-called „C-Scan"?
15. Name some of the typical structural components, which are commonly produced by prepreg autoclave technology.
16. Where do you see potential for improvement for prepreg autoclave technology?
17. Name at least two different infusion processes used for the production of structural airframe components. What are the differences?
18. Why are heavy tools and special clamping devices required for the classic RTM process?
19. How can textiles be fixed, that shall be processed by means of resin infusion?
20. What needs to be considered with respect to the infusion temperature of the resin?
21. What are some of the advantages of the textile/infusion technique over the prepreg autoclave technology?
22. What are some of the disadvantages of the textile/infusion technique over the prepreg autoclave technology?
23. What are some examples of structural components, which are produced by textile/infusion technique and used in aircraft construction?
24. How are structural components usually manufactured in thermoplastic technology for commercial aircraft?
25. What thermoplastic matrix systems are normally used for structural components in commercial aircraft?
26. What advantages does the thermoplastic technology provide for the production of structural components?
27. What disadvantages exist concerning thermoplastic technology when producing structural components?
28. What is the decisive deformation mechanism in forming fabric-reinforced thermoplastics, and what has to be considered concerning the tool design?
29. What are benefits expected from thermoplastic tape laying?
30. For which components is the filament winding technology used?
31. What is the dominating joining method for CFRP airframe structures?
32. Identify advantages of riveting!
33. What are disadvantages of blind bolts?
34. What are disadvantages of riveted joints?
35. Which structural bonding method is mostly used for secondary bonding?
36. How can a CFRP adhesive surface be prepared for structural bonding?
37. What is a „water break test"?

38. What difficulty exists regarding the quality assurance of structural adhesive bonding?
39. Which non-destructive test method is used mainly for bonded CFRP primary structures, and what exactly can be made visible with it?
40. What joining methods are commonly used for the assembly of CFRP components (wings, tail, fuselage barrels, etc.)?

Exercise: Manufacturing Technology Selection and Assessment

1. CFRP skins, stringer, frames and clips shall be manufactured for a fuselage barrel. Only monolithic design solutions (no sandwich solutions) are considered. Prepare simplified sketches of the main process steps of the following manufacturing process technologies: ATP, AFP, MVI, Pultrusion and Thermoplastic Stamp Forming.
2. Estimate (approximately) the part dimensions (length, thickness, width) of skins, stringer, frames and clips for a single aisle aircraft fuselage barrel.
3. Prepare a catalogue with important criteria for the selection and evaluation of the different manufacturing process technologies for the different parts. Concentrate on criteria for optimised part properties during the operational phase and on manufacturing cost relevant criteria.
4. Prepare a quantitative assessment according to Table 4.3 and provide rationale for your final decision on the favourable manufacturing process technologies for skins, stringer, frames and clips.

References

1. Aerospace Manufacturing, vol. 7, Issue 62, p. 50 (October 2012)
2. Aerospace Manufacturing, vol. 9, Issue 78, pp. 32–33 (May 2014)
3. Petermann, J.: Automatisierungsbedarfe in der CFK-Produktion, 2. Augsburger Produktionstechnik-Kolloquium 2013, DLR ZLP, Augsburg 15.5.2013
4. mTorres. http://www.mtorres.es/en/aeronautics/products/carbon-fiber/torreslayup
5. Flemming, M., Ziegmann, G., Roth, S.: Faserverbundbauweisen, Fertigungsverfahren, Fertigungsverfahren mit duroplastischer Matrix, Auflage: 1. Springer, Berlin (1998). ISBN-13: 978-3540616597
6. Breuer, U.: Beitrag zur Umformtechnik gewebeverstärkter Thermoplaste, VDI Fortschritt-Berichte Nr. 433, Reihe 2: Fertigungstechnik. VDI Verlag, Düsseldorf (1997)
7. Fivesgroup. http://metal-cutting-composites.fivesgroup.com/
8. Breuer, U.: Herausforderungen an die CFK Forschung aus Sicht der Verkehrsflugzeug Entwicklung und Fertigung, 10. Nationales Symposium SAMPE Deutschland e.V., Darmstadt, 2005
9. Jamco. www.jamco.co.jp/e/jco/adp.html
10. Spengler, M.: Automation der PAG A350 Schalenproduktion, 3. Augsburger Produktionstechnik-Kolloquium, DLR ZLP, Augsburg, 19.5.2015

11. Ablitz, D., et al.: Curing methods for advanced polymer composites – a review. Polym. Polym. Compos. **21**(6), 341–348 (2013)
12. v. Wachter, F.K.: Ein Beitrag zur zerstörungsfreien Ultraschallprüfung von Faserverbundwerkstoffen, pp. 22 and 46. Dissertation RWTH Aachen, Verlag Dipl.-Ing. Chaled Shaker, Aachen (1992)
13. Netzsch Analysing & Testing: Dielectric Cure Monitoring, DEA288 Epsilon Product Information. https://www.netzsch-thermal-analysis.com/en/products-solutions/dielectric-analysis/
14. Knappe, S., Schmölzer, S., Küchenmeister-Lehrheuer, C., Oldörp, K., Barber, K.: Simultaneous Rheological-Dielectrical Characterisation using the HAAKE MARS Rheometer Platform. Thermo Fisher Product Information P055
15. Olympus NDT: Introduction to Phased Array Ultrasonic Technology Applications, 3rd edn. (2007). ISBN 0-9735933-4-2
16. Olympus NDT: Phased Array Testing – Basic Theory for Industrial Applications, 2nd edn. (2012)
17. Olympus NDT: Ultrasonic Transducers – Wedges, Cables, Test Blocks. www.olympus-ims. com
18. Schiebold, K.: Zerstörungsfreie Werkstoffprüfung – Ultraschallprüfung. Springer, Berlin (2015). ISBN 978-3-662-44699-7
19. Breuer, U.: Reinforcement of CFRP structures by tailored fibre placement. Polym. Polym. Compos. (GB) **6**(8), 499–504 (1998)
20. Becker, D.: Transversales Imprägnierverhalten textiler Faserstrukturen für Faser-Kunststoff-Verbunde. IVW Verlag, Kaiserslautern (2015)
21. Neitzel, M., Mitschang, P., Breuer, U.: Handbuch Verbundwerkstoffe. Carl Hanser Verlag, München (2014). ISBN 978-3-446-43696-1
22. Potter, K.: Resin Transfer Moulding. Chapman & Hall, London (1997). ISBN 0412725703
23. http://composyt.com/vap-en/
24. Filsinger, J., et al.: Method and device for producing fiber-reinforced components using an injection method. US Patent 6 843 953 B2, 18 Jan 2005
25. Filsinger, J: New variant of VAP (Vacuum Assisted Process) to enable an automated process setup. SAMPE Conference Proceedings, Long Beach, California, 6–9 May 2013
26. Deutsches Patent- und Markenamt Offenlegungsschrift DE 10 2010 025 068
27. United States Patent Application Publication US 2013/0099429, 25 Apr 2013
28. Filsinger, J.: Deutsches Patent-und Markenamt Offenlegungsschrift DE 10 2013 006 940 A1
29. International Patent Application WO2014/173389 A1, 30 Oct 2014
30. Hexcel HexFlow® RTM6 Product Data Sheet. http://www.hexcel.com/Resources/rtm-data-sheets
31. Schnitzer, M.: Anforderungen und Lösungsansätze für einen höheren Automatisierungsgrad in der CFK Fertigung, 2. Augsburger Produktionstechnik-Kolloquium DLR ZLP, Augsburg 15.5.2013
32. Neise, E.: Processing of Fibre Reinforced Polymers using Industrial Robots. Dissertation Rheinisch-Wetsfälische Technische Hochschule Aachen, Germany (1986)
33. SQRTM enables net-shape parts. In: High Performance Composites, September 2010. http://www.compositesworld.com/articles/sqrtm-enables-net-shape-parts
34. Gueuning, D., Mathieu, F.: Evolution in composite injection moulding processing for wing control surfaces. SAMPE Conference Proceedings, Sampe Europe 2015, Amiens, France, 15–17 Sept 2015
35. http://cordis.europa.eu/project/rcn/52465_en.html
36. http://www.hexcel.com/products/industries/imultiaxials-nc2
37. Friedrich, K., Breuer, U.: Multifunctionality of polymer composites – challenges and new solutions. Elsevier (2015). ISBN 978-0-323-26434-1
38. Cytec Engineered Materials Technical Data Sheet: CYCOM® 977-20 RESIN SYSTEM WITH PRIFORM™ TECHNOLOGY. https://www.cytec.com/

39. Breuer, U., et al.: Deep drawing of fabric reinforced thermoplastics. Polym. Compos. (USA) **17**(4), 643–647 (1996)
40. Breuer, U., Neitzel, M.: High speed stamp forming of thermoplastic composite sheets. Polym. Polym. Compos. (GB) **4**(2) (1996)
41. Breuer, U., Ostgathe, M.: Manufacturing of all-thermoplastic sandwich systems by a one-step forming technique. Polym. Compos. (USA) **19**(3), 275–279 (1998)
42. Duhovic, M., Mitschang, P., Bhattacharyya, D.: Modelling approach fort the prediction of stitch influence during woven fabric draping. Compos. A: Appl. Sci. Manuf. **42**(8), 968–978 (2011). http://www.sciencedirect.com/science/article/pii/S1359835X11001059
43. Schommer, D., Duhovic, M., Hausmann, J.: Modeling of non-isothermal thermoforming of fabric reinforced thermoplastic composites. Proceedings 10th European LS-Dyna Conference 2015, Würzburg, Germany. https://www.dynamore.de/de/download/papers/2015-ls-dyna-europ/documents/sessions-d-1-4/modeling-non-isothermal-thermoforming-of-fabric-reinforced-thermoplastic-composites
44. Offringa, A.: Redesigned A340-500/-600 fixed wing leading edge (J-Nose) in thermoplastics. SAMPE Europe Conference Proceedings 2001, Paris
45. Thermoplastic composites gain leading edge on the A380. In: Composites World, High Performance Composites (2006). http://www.compositesworld.com/articles/thermoplastic-composites-gain-leading-edge-on-the-a380
46. Miaris, A., Edelmann, K., Sperling, S.: Thermoplastic matrix composites: Xtra complex, Xtra quick, Xtra efficient-manufacturing advanced composites for the A350 XWB and beyond. Proceedings 20th International Conference on Composite Materials, Copenhagen, 19–24 July 2015
47. Black, S.: Thermoplastic composites "clip" time, labor on small but crucial parts. Compos. World **1**, 66 (2015)
48. Edelmann, K.: CFK-Thermoplast-Fertigung für den A350XWB, lightweightdesign 2/12
49. FLUG REVUE 07/2014
50. Offringa, A.: Development of an aircraft torsion box with an integrally stiffened thermoplastic skin. SAMPE Japan (JISSE-12) Conference Proceedings, Tokyo, 9–11 Nov 2011
51. Doldersum, M.H.J.: Industrialization of thermoplastic control surfaces. Conference Proceedings SAMPE Europe's 35th International Conference (SEICO 14)
52. Offringa, A., Davies, C.R.: Gulfstream V floors – primary aircraft structure in advanced thermoplastics. J. Adv. Mater. **27**(2), 2–10 (1996)
53. Offringa, A., List, J.: Lightweight thermoplastic beams. SAMPE Conference Proceedings, Dallas (2006)
54. Beresheim, G.: Thermoplast-Tapelegen – ganzheitliche Prozessanalyse und –entwicklung, IVW Schriftenreihe Band 32. Institut für Verbundwerkstoffe IVW GmbH, Kaiserslautern (2002). ISBN -3-934930-28-X, ISSN-1615-021X
55. Khan, M.A.: Experimental and Simulative Description of the Thermoplastic Tape Placement Process with Online Consolidation, IVW Schriftenreihe Band 94. Kaiserslautern (2010). ISBN 078-3-934930-90-2, ISSN 1615-021X
56. Jäschke, A.: Funktionaler Leichtbau durch Spriform, 35. Internationaler Kongress – Kunststoffe im Automobilbau, Mannheim, 6.-7. April 2011
57. Moser, L.: Experimental Analysis and Modeling of Susceptorless Induction Welding of High Performance Thermoplastic Polymer Composites. IVW Schriftenreihe, IVW Kaiserslautern (2012)
58. Funck, R.: Entwicklung innovativer Fertigungstechniken zur Verarbeitung kontinuierlich faserverstärkter Thermoplaste im Wickelverfahren, VDI Berichte Reihe 2 Nr. 393. VDI-Verlag, Düsseldorf (1996). ISBN 3-18-339302-6
59. Miaris, A.: Experimental and Simulative Analysis of the Impregnation Mechanics of Endless Fibre Rovings, IVW Schriftenreihe Band 102. Institut für Verbundwerkstoffe GmbH, Kaiserslautern (2012). ISBN 978-3-934930-98-8, ISSN 1615-021X

60. Bautz, B., et al.: INCCA „Integration von hybriden Zug-Crashabsorbern im Flugzeug-Cargobereich" Luftfahrt-Forschungsprogramm Abschlussbericht Verbund TENOR, Förderkennzeichen 20W1105D, Laufzeit des Vorhabens, 01.01.2012–31.03.2015

61. Kopietz, M., Grishchuk, S., Wetzel, B.: Toughening of thermosetting cyanate ester hybrid resins with commercially available nanodispersed SiO2, Abstract Book, p. 102. International Research and Practice Conference "Nanotechnology and Nanomaterials" (NANO-2015), Lviv, Ukraine, 26–29 Aug 2015

62. Hi-Shear Corporation, 2600 Skypark Drive, Torrance, CA 90509, U.S.A., www.hi-shear.com, HI-LOK^{TM} fasteners available to the aerospace industry are listed on the LISI AEROSPACE webpage http://www.lisi-aerospace.com/products/fasteners

63. Henning, F., Möller, E.: Handbuch Leichtbau, p. 763. Carl Hanser, München (2011). ISBN 978-3-446-42267-4

64. Brötje: http://www.broetje-automation.de/loesungen-und-kundennutzen/equipment/auto mated-assembly/drilling-fastening/isac

65. Geiß, P.: Oberflächentechnologie, Scriptum zur Vorlesung 2013/2014, Universität Kaiserslautern. http://www.mv.uni-kl.de/awok/home/

66. Geiß, P.: Fügeverfahren für Verbundwerkstoffe, Scriptum zur Vorlesung 2013/2014, Universität Kaiserslautern. http://www.mv.uni-kl.de/awok/home/

67. Holtmannspötter, J.: Strukturelles Kleben von CFK Strukturen, Wehrwissenschaftlichen Institut für Werk- und Betriebsstoffe. Proceedings DGLR Congress, Stuttgart, 10.-12.9.2013

68. Wolf, M.: Oberflächeninspektion und Oberflächenvorbehandlung, IFAM Bremen. Proceedings DGLR Congress, Stuttgart, 10.-12.9.2013

69. Dreyer, C.: Fraunhofer PYCO Polymermaterialien und Composite „Wiederlösbares Kleben – ein neues Klebkonzept". Proceedings Bonding Visions, Hamburg, 15.-16.5.2014

70. Sorochynska, L., Motsch, M., Magin, M.: Thermoplast-Duroplast Verbindungen – Einfluss der Funktionsschicht auf mechanische Eigenschaften, 3. Sitzung DGM-Fachausschuss „Hybride Werkstoffe und Strukturen", Kaiserslautern, 9. Oktober 2013

Chapter 5
Testing

Abstract In order to certify the aircraft, the structural integrity and the safety of the airframe structure are usually demonstrated by analysis and tests. In addition, structural tests are also necessary to verify new design methods and manufacturing technologies (including acceptable tolerances of defects), to assess new materials and joining methods and for many other development and certification purposes described in this chapter. The building block approach of the test pyramid is discussed, providing examples for each level of the pyramid, i.e. coupons, elements, details, sub-components, components and major tests. Finally, a possible major test procedure for compliance demonstration is described.

Keywords Validation and verification • Test pyramid • Building block • Coupon test • Element test • Detail test • Sub-component test • Component test • Major test • Compliance demonstration • Generic test • Specific test • Development test • Tear down

Validation and Verification

The structural integrity and the safety of the airframe are demonstrated by analysis and tests. Analysis and test evidence is presented to the airworthiness authorities to demonstrate compliance with the certification rules. In Europe, the "Certification Specifications for Large Aeroplanes (CS-25)" edited by the European Aviation Safety Agency EASA is applicable [1].

A process often referred to in airframe development is the "V&V" process: Validation and verification. "*Validation*" covers all activities during the design phase needed to agree on requirements and to assess that they have been properly introduced to the design, see Chap. 2.

"*Verification*" of a structural design means checking that the manufactured structure meets the requirements by structural testing, i.e. that it sustains the relevant mechanical and thermal loads.

© Springer International Publishing Switzerland 2016
U.P. Breuer, *Commercial Aircraft Composite Technology*,
DOI 10.1007/978-3-319-31918-6_5

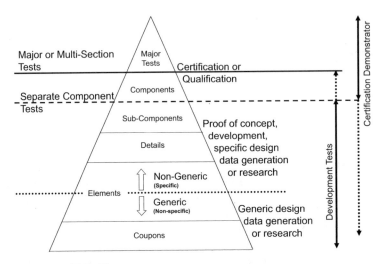

Fig. 5.1 Test pyramid [1, 2]

Test Pyramid

The usual test approach is based on the test pyramid, Fig. 5.1. The test pyramid is established in order to minimise time and cost effort while ensuring that all needed test results during the development and certification phase are available in time and in the quantity and quality needed. Test sample size, its complexity, manufacturing lead time and cost increase from bottom to top of the pyramid.

In the development phase of the aircraft, each critical loading condition of the structure is analysed to demonstrate compliance with the strength and deformation requirements. This includes comprehensive assessment of the external loads, the resulting internal strains and stresses and the structural allowables, see Chap. 2, Fig. 2.10. The certification rules provide the relevant information on conditions and specifications (e.g. ground and flight manoeuvres, load factors, safety factors, etc.) that must be applied; they also provide some detailed guidance for the kind of testing that is typically required for compliance demonstration.

Even during the concept phase, before the "authorisation to offer" (see Chap. 2, Fig. 2.1), all certification risks linked to the new aircraft should be carefully analysed and assessed in terms of potential mitigations and airworthiness demonstration, and this should be done in close cooperation with representatives from the airworthiness authorities. For this purpose, integrated teams with engineers from R&T, stress, materials, manufacturing, in-service support, systems, flight physics and from other disciplines have to analyse the key differences from existing aircraft, which might have an impact on safety and airworthiness demonstration, for example

- type of engine and engine position
- landing gear architecture and position

- specific wing, fuselage or tail plane architecture
- high lift configuration
- *new materials*
- other differentiating technologies

During the development phase, a certification plan is established in cooperation with representatives from the airworthiness authorities, including a list of all certification tests. Compliance with the rules has to be demonstrated to obtain the type certification. Due to the fact that several tests have to be performed during the flight test phase, certain *major tests* (i.e. at the top of the test pyramid, Fig. 5.1) on ground are necessary to get clearance for the first flight. These tests include for example (list not exhaustive):

- 1 g bending test of wings (checking for example gaps and interference of flaps and slats)
- maximum wing bending at limit load condition
- functioning of flight controls (aileron, spoiler) under limit load

To receive the type certificate (and to be able to deliver the first aircraft to the first customer airline for commercial operation), the following structural major tests have to be carried out (list not exhaustive):

- ultimate load tests
- residual strength tests

Besides certification, structural tests are also required for development purposes, for example (list not exhaustive):

- analysis method verification (including FEM validation)
- performance assessments of new materials and joining technology
- verification of new design methods
- verification of new manufacturing methods
- verification of concession limits for manufacturing quality findings (i.e. assessment of acceptable tolerances of manufacturing defects)
- investigation of durability
- investigation of maintainability
- verification of (new) repair schemes
- verification of virtual testing methods
- other

Structural tests are also necessary for the development and verification of in-service support methods.

Finally, structural tests are performed for research purposes, prior to the concept phase of a new aircraft, in order to assess new technology solutions in terms of weight, cost and risk.

Figure 5.2 provides an overview of some typical tests which are performed during the development phase of a new aircraft.

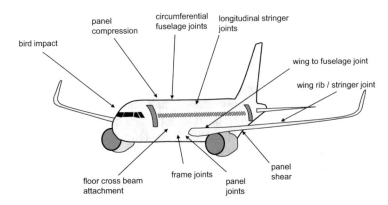

Fig. 5.2 Overview of some typical structural tests (not exhaustive)

The test pyramid starts with *coupons* [2]. Coupons are relatively small test specimen (typically only a few cm wide and long, and only a few mm thick, usually flat laminates), the precise dimensions are part of the relevant test specification which can be used to evaluate a material. Typical CFRP coupon tests are for example (not exhaustive):

- water uptake
- determination of glass transition temperature T_G by DSC, DMTA for dry and h/w condition
- tension (strength and stiffness, plain and with open or filled holes, also after exposure to aggressive media and under elevated and lowered temperatures)
- compression (as above)
- interlaminar shear strength
- lap shear strength of bolted or bonded joints
- pin loaded bearing strength (bolted joints)
- G_{ic} (determination of fracture toughness)
- impact tests, incl. the determination of outer dent depth and inner damages at different impact energy levels, and including the determination of residual compression strength after impact

Definitions, recommendations and examples for tests to be carried out for different purposes can be found in CS-25 Book 2, AMC subpart D [2]. A description of important CFRP coupon test procedures can be found in [3] and [4].

The next stage of the pyramid is defined by *elements*. Elements can be specific (non-generic), i.e. representing a part of a structural design solution of an airframe, with the same function and with the same sizing. Elements can also be non-specific (generic). This indicates that the test specimen does not represent the specific function and sizing associated with a particular aircraft project.

Typical CFRP element tests are for example (not exhaustive):

- skin lightning strike behaviour
- stiffener compression and tension behaviour

- skin/stringer joint impact resistance
- sandwich panel strength and stiffness

A *detail* is a specific, non-generic and more complex airframe structure, such as for example:

- longitudinal or circumferential fuselage panel joints
- stiffened panels with specific stringer run-outs at loaded or unloaded cut-outs

Figure 5.3 shows the test set up of a CFRP compression panel. The image was taken after the stability limit was reached, and the large sideward deflection of the three stiffeners is clearly visible. The stability limit can be assessed by means of strain gauges mounted on the front and the rear side of the panel. The two curves split when the skin starts to buckle; from this point on the strain gradient increases on the convex side and decreases on the concave side of the panel skin.

Other examples for panel tests are examinations to evaluate the flammability, smoke and toxicity behaviour of the structure. Burn-through tests are usually made with compression panels, simulating an engine fire, in which the CFRP panel surface is directly exposed to flame temperatures of more than 1000 °C for several minutes, and the residual compression strength is determined in a subsequent test.

A *sub-component* is a major airframe structure which can provide structural representation of a section of the complete structure. Typical examples:

- wing stub-box
- HTP or VTP stub-box
- fuselage or wing panel
- frames

A *component* is a major section of the airframe, such as

- wing
- fuselage
- HTP, VTP
- landing gear

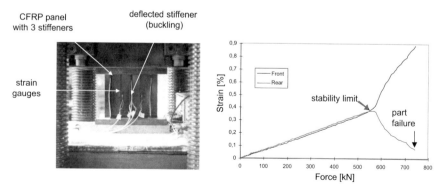

Fig. 5.3 CFRP compression panel test set up and "chute"-curve of measured strain (front and rear mounted strain gauges)

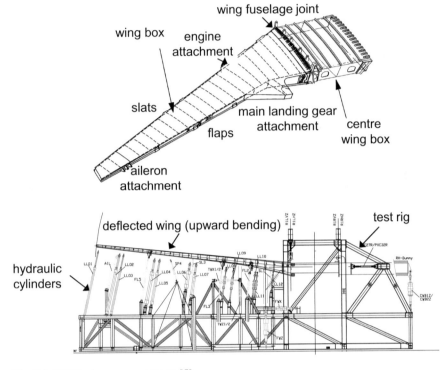

Fig. 5.4 CFRP wing component test [5]

Figure 5.4 shows the test set up of a CFRP wing, which was developed within an R&D program funded by the German government [5]. The wing was attached to a centre wing box in order to ensure a proper load introduction. This wing with a half span of 15.58 m was designed for a 100 passenger aircraft with an MTOW of 46.5 t. The test program was very extensive and comprised—among other tests—ground vibration tests, leak tests, limit load test, fatigue test (60,000 simulated flights), ultimate load test, a damage tolerance test (after artificial damages were introduced) with 30,000 simulated flights, and a residual strength test. The wing was filled with water, simulating the kerosene.

Finally, *major tests* or *multi-section tests* form the top of the test pyramid. This means tests of a full scale aircraft airframe, comprising two or more components.

Figure 5.5 gives an impression of the A380 major test. The hanging metal profiles are used for load introduction into the wing structure; the attachment points on top of the wing are clearly visible. A test rig for a complete aircraft can contain hundreds of jacks and load introduction devices, and thousands of gauges and cables. The weight of the rig can be up to 1000 t or more. It is obvious that the design of the test rig for a major test is time consuming, and its manufacturing and installation is very expensive. This underlines the necessity to answer as many questions as possible by test results gained from the lowest possible level of the test

Fig. 5.5 A380 major test [6]

pyramid. The test campaigns with major test components before type certification and entry into service usually run for more than 1 year. A possible procedure to demonstrate compliance is

- manufacturing of the major components with maximum allowed manufacturing defects (in case of CFRP for example delamination, weak or missing bond lines, inclusion of foreign objects etc., refer to Chap. 3) and structural repairs
- introduction of maximum allowable (non-visible) impact damages
- dynamic loading of the structure to simulate flight and ground loads of a complete design service goal
- for the dynamic loading, a load elevation factor to compensate for any material strength reduction due to temperature and humidity can be applied; this load elevation factor can be defined by the environment knock down factor EKDF, which for CFRP can typically vary between 10 and 20 % (i.e. in this case the load elevation factor could be between 1.1 and 1.2)
- an ultimate load test is performed (in which the structure must not fail)
- visible impact damages and large damages are introduced
- dynamic loading of the structure to simulate flight and ground loads of half the design service goal
- limit load is applied (and the structure must not fail)
- a residual strength test is performed

In many cases, "tear down" programs are performed after the residual strength test, in which the component is loaded until it fails. The tested component is disassembled and individual parts are carefully examined.

Questions

1. Related to design and test of airframe structures, what does "V&V" mean?
2. What are the main activities during the "V&V"-process?
3. What are structure tests needed for?
4. What is a test pyramid? What is its purpose?
5. Give examples for the different stages of a test pyramid.
6. What kinds of tests are performed?
7. Give examples for CFRP coupon tests.
8. What does "major test" mean?
9. What does "specific test" and what "non-specific test" mean?
10. Why is the influence of moisture and temperature on CFRP properties usually investigated on the lowest level of the test pyramid?
11. How would you take the influence of moisture and temperature on CFRP into account during a component test?
12. Are major tests necessary before first test flights?
13. How can the damage tolerance of a CFRP structure be investigated during a component test?

References

1. Certification Specifications for Large Aeroplanes (CS-25). https://easa.europa.eu/certification-specifications/cs-25-large-aeroplanes
2. Certification Specifications for Large Aeroplanes (CS-25), Book 2, AMC No.1 to CS 25.603, Subpart D
3. Adams, D.F., Carlsson, L.A., Pipes, R.B.: Experimental Characterization of Advanced Composite Materials, 3rd edn. CRC Press, Boca Raton, FL (2002). ISBN 9781587161001
4. Grellmann, W., Seidler, S.: Kunststoffprüfung, 2. Auflage. Carl Hanser Verlag, München (2011). ISBN 978-3-446-42722-8
5. CFK-Tragflügel Abschlussbericht 1.9.1997-30.09.1999, Bundesministerium für Bildung und Forschung, BMBF Förderkennzeichen 20W 9703 A
6. Orsenna, É.: A380, Fayard Editions, p. 102. Xavier Barral, Paris (2007)

Chapter 6
Repair

Abstract A maintenance program is mandatory for commercial aircraft; thus its definition must be part of the aircraft development. This chapter describes the basic parts of the maintenance program and provides examples for interval and duration of scheduled maintenance (checks). Examples of structural damages and principles of the structural repair manual (SRM) are explained, including the definitions of cosmetic, structural, temporary and permanent repair. Important accidental damages are introduced, highlighting the necessity of a robust, impact-resistant yet not too heavy CFRP design. The detectability threshold of CFRP impact damages, crucial to decide on cosmetic or structural repair, is explained. The rationale of bolted repair and a possible repair procedure are described. Advantages and challenges of adhesive bonding repair solutions are discussed, highlighting the specific restrictions of airworthiness authorities.

Keywords Maintenance program • Aircraft maintenance manual • Structural repair manual • Inspection interval • A-check • B-check • C-check • D-check • Overhaul • Damage description • Cosmetic repair • Structural repair • Temporary repair • Permanent repair • Damage size classification • Accidental damage • Impact damage event • Indentation depth • Detectability threshold • Barely visible impact damage BVID • Residual strength • Bolted repair • Bypass failure • Bearing failure • Shear failure • Cleavage failure • Bolt failure • Counter sunk rivet • Minimum skin thickness • Bonded repair • Surface contamination • X-ray fluorescence analysis • Infrared spectroscopy • Weak bond • Kissing bond

Maintenance Program

As stated in Chap. 5, a CFRP airframe structure has to fulfil all certification relevant requirements. For large commercial aircraft the "Certification Specifications for Large Aeroplanes (CS-25)", edited by the European Aviation Safety Agency EASA, is applicable. However, on top of this, aircraft operators want a maximum of robust and damage tolerant structures in order to maintain aircraft availability and to avoid repair and maintenance effort as much as possible. Trading structure light weight efficiency (thinner structures = less fuel burn cost) versus structure robustness (thicker structure = less maintenance cost), the CFRP design must be

© Springer International Publishing Switzerland 2016
U.P. Breuer, *Commercial Aircraft Composite Technology*,
DOI 10.1007/978-3-319-31918-6_6

optimised to provide enough robustness without too much penalising the light weight performance. Repair methods must be available, including cosmetic repair (for optical reasons) and structural repair to recover the load carrying capability of the relevant part, if this becomes necessary as a consequence of damage.

AMC No. 1 to CS-25.603 §8.8 "Substantiation of Repair" requires *"When repair procedures are provided in maintenance documentation, it should be demonstrated by analysis and/or test, that methods and techniques of repair will restore the structure to an airworthy condition."* [1]

The substantiation of repairs is done by analysis, which is supported by tests (Test Pyramid Building Block Approach, see Chap. 5): From coupons to elements, tests are dedicated to substantiate repair solutions. Analysis is performed to demonstrate the structural repair capability. Representative damage sizes are introduced in full scale test specimens to prove the maximum allowable damage size for each part of the structure. Full scale test specimens also include representative repair solutions. The repaired structure has to fulfil the same requirements (i.e. restoring ultimate load capability) as the parent structure.

CS 25.1529 "Instructions for Continued Airworthiness" require the existence of a maintenance program. Figure 6.1 provides an overview of the different tasks and responsibilities of aircraft manufacturer, operator, maintenance shop (in some cases, operators task certified maintenance shops with repair and maintenance of their aircraft) and authorities.

The maintenance program is developed prior to type certification.

It usually contains (list not exhaustive):

- a description of the location to be covered by an inspection task
- the definition of the level of inspection (e.g. general visual inspection, detailed visual inspection, special and detailed inspection, servicing, functional check etc.)
- a consideration of accidental damage

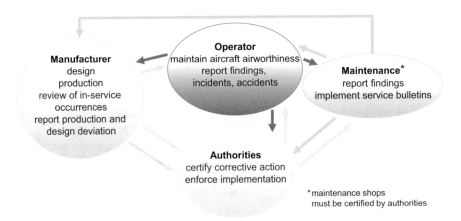

Fig. 6.1 Maintenance program tasks and responsibilities

The related inspections must be included within the aircraft justification documentation for type certification. After type certification, the maintenance program is further developed in a continuous process throughout the operational life of the aircraft to ensure its airworthiness. This is achieved by (list not exhaustive):

- revisions following the full scale fatigue test results
- revisions following in-service occurrence (e.g. after lightning strike, hard landing, bird strike, tail strike, ...)
- adjustments following in-service feedback (from service reports, surveys, measurements, ...)

The documentation can be structured as follows:

A "Maintenance Planning Document" (MPD) can contain the summary of all tasks and build the basis for the maintenance programme of the operating airline. It has to be agreed with the airworthiness authorities.

The "Aircraft Maintenance Manual" (AMM) contains detailed instructions for scheduled maintenance for structure, equipment and systems [2, 3]:

- *A-Check:* Typically every 1–2 months (usually after 250–650 flight hours, depending on the type of aircraft), usually done within one night, routine check making sure everything is functioning safely and efficiently
- *B-Check:* Typically every 3–5 months (usually after every 1000 flight hours), usually this check takes 0.5–1 day (about 150 man hours), and is more extensive than A-checks
- *C-Check:* Typically every 1–1.5 years, the check usually takes 1–2 weeks and requires the aircraft to be docked at a hangar or repair station for detailed inspections
- *D-Check:* Usually every 4–10 years, takes approx. 4–6 weeks (typically 55,000 man hours for a B747-400); it is a general overhaul, with large parts of the aircraft being dismantled, carefully inspected and re-assembled

There is a clear tendency to extend the intervals (the newer the aircraft program the longer the interval), and at the same time to reduce the "lay down time" of the aircraft during the checks by spreading tasks. A D-check is scheduled after 12 years for A350XWB.

The "Structural Repair Manual" (SRM) describes the structure, repair materials, tools and repair processes. It also defines allowable damage limits (ADL). This is the limit below which the structure loses its ultimate load carrying capability. SRM repairs are fully validated and verified, and can be used by any qualified MRO maintenance shop (MRO = maintenance, repair and overhaul). Typically the damages are relatively small, and the structure load carrying capability is not below limit load when left unrepaired. In other cases, the aircraft manufacturer must be involved.

For a repair of a structure, the basic steps are described in Fig. 6.2, [4]. For the ease of its usage, the SRM follows the order of the International Air Transport Association (ATA) Specification 100 [5].

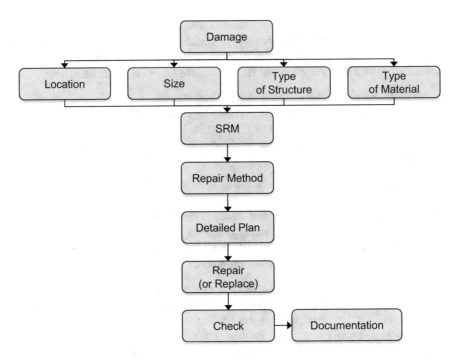

Fig. 6.2 Basic steps of a repair procedure. Flowchart based on [4]

The aircraft manufacturer can define "restricted areas", for example classified with the capital letter A "critical", B for "semi-critical", or C for "uncritical", in order to indicate where repairs must not be carried out without the involvement and agreement of the manufacturer or a certified repair station. Typically this is the case in areas which are very sensitive to the aerodynamic performance of the aircraft (wing leading edges), where important probes are positioned (forward fuselage) or at other specific areas such as cargo door corners, flap track beams, flaps and ailerons.

In order to assess a damage and to decide on the proper way of repair, the SRM contains very detailed descriptions and definitions for all kinds of damages of the relevant structure (ATA chapter, Table 6.1) such as scratches, gouges, marks, cracks, dents, nicks, distortion, corrosion, creases, abrasion, debonding, delamination, fretting, indentation, burn marks, penetration and other damages. A few examples are shown in Table 6.2.

Cosmetic and Structural Repair

Cosmetic repairs must be distinguished from structural repairs, and temporary repairs must be distinguished from permanent repairs.

Table 6.1 ATA classification (excerpt) [5]

Title	Chapter
Structure	50
General	51
Doors	52
Fuselage	53
Nacelles and pylons	54
Stabilizers	55
Windows	56
Wings	57

Table 6.2 Damage description (examples) [4]

Designation	Explanation	Symbol
Dent	Soft edges, no penetration	
Crease	Sharp edges, no penetration	
Abrasion	Loss of thickness	
Gouge	Loss of thickness	
Nick	Close to edges	
Scratch	Sharp, tapered	
Fracture	Through thickness damage	
Corrosion	Reduction of thickness	

Cosmetic Repair means:

- a repair is not necessary from a structural (load carrying and safety) point of view
- although a damage was detected, the structure still has its full ultimate load capability
- a repair is not mandatory to maintain airworthiness
- the surface of the damaged structure can be restored by the operator (or the maintenance shop) for optical reasons if he wishes to do so

Structural Repair means:

- repair is mandatory
- full ultimate load capability will be restored by the repair
- SRM instructions must be applied
- in cooperation with or by the aircraft manufacturer a large repair can be performed and validated (outside of the SRM), for example a repair of a damage that does not allow continuation of operation with less than limit load capability

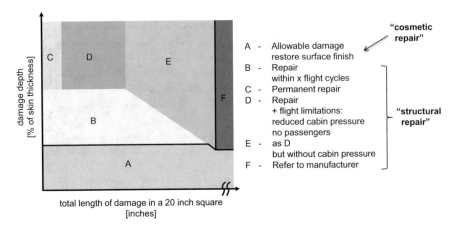

Fig. 6.3   Damage size classification [4]

Temporary Repair means:

- a repair can be made in non-restricted areas
- full load capability will be restored
- the repair can enable only a limited number of flight cycles and/or flight hours

Permanent Repair means:

- a repair is made in non-restricted areas and restricted areas (with manufacturer permission)
- full load capability will be restored for the complete design service goal

In order to assess a given damage and to decide on the necessary repair action, the SRM provides additional information related to location, size and severity of damages. A very simplified example, just to demonstrate the principle, is shown in Fig. 6.3.

For very small damage depths (e.g. scratches only within the paint), a cosmetic repair solution can be applied. Once the depth exceeds a certain value, a structural repair is necessary to maintain airworthiness.

There are many different origins of accidental damages, for example

- runway debris
- hailstorm
- lightning strike
- tail strike
- bird strike
- foreign object damages during cargo loading or unloading
- maintenance (tool drops, accidental coating removal, ...)
- ...

Impressive images of severe tail strike, lightning strike and bird strike damages can be easily found in the internet.

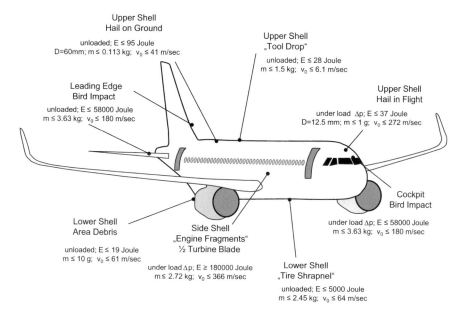

Fig. 6.4 Impact events (examples only, values not to be taken for design, based on [7, 8])

In [6] it is reported that Lufthansa Technik recorded more than 1647 accidental damage events for 243 of its aircraft in 2006, and among these were 1248 collisions, 216 lightning strikes and 184 bird strikes. A damage event at the fuselage was recorded approximately every 3000 flights. The repair cost was above 33 M€. This underlines the necessity of highly damage tolerant material and design solutions especially for the fuselage structure.

The CFRP fuselage is relatively new in commercial transportation (B787 entry into service 25th September 2011, A350XWB 18th December 2014), and has yet to prove its in-service capabilities, however, a lot has been done to provide damage tolerance. The developments go back to the 1990s. Figure 6.4 shows assumptions for fuselage impact events which have been used for the German R&D project "Schwarzer Rumpf" by Hachenberg [7].

Impact is particularly important for CFRP, since certain impact events damages within the laminate (fibre and matrix failure, delamination, cracks) cannot be seen from outside, as the surface seems to be intact. A cross section view of an impact damage of a CFRP laminate shows the typical "fir tree pattern", Fig. 6.5, i.e. the damage spreads from top to bottom of the CFRP laminate. This is schematically illustrated in Fig. 6.6.

Having clear decision criteria for cosmetic or structural repair is crucial. The dent depth can be used for this purpose; however, this means that any internal damages caused by impact which are below the detectability threshold must be sustained by the structure, and ultimate load capability must be demonstrated by

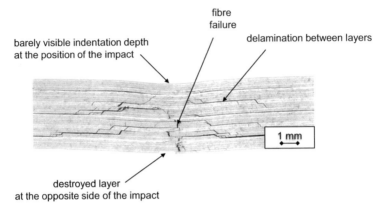

Fig. 6.5 Cross section of an impacted CFRP panel with the typical fir tree pattern of the damage

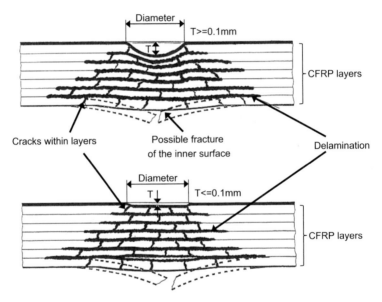

Fig. 6.6 CFRP laminate cross section after impact (schematically) with barely visible indentation depth (*top*) and non-visible indentation depth (*bottom*)

appropriate design (with allowables experimentally gained from damaged structures) and test.

Figure 6.7 shows the residual strength of a CFRP structure as a function of impact energy. For small impact energies, damages will not occur (no visibility from outside, and no inner visibility, either), and the residual strength will be at the level of the original strength. If the impact energy is increased, damages can occur inside the CFRP laminate ("inner visibility") which are, however, not visible from outside, and the residual strength of the structure will decrease. At a certain impact

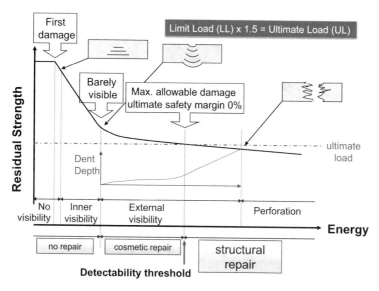

Fig. 6.7 CFRP residual strength as a function of impact energy (based on [9], for further information see also [10])

energy level, permanent dents within the CFRP surface will occur ("external visibility"). As long as the residual strength capability of the CFRP structure is above ultimate load, these impact damages are classified as "barely visible", and no structural repair is required. Cosmetic repair can be applied but is not mandatory. The detectability threshold is reached at impact energy levels where the residual strength of the structure falls below ultimate load. At this point, it must be ensured and verified by appropriate design and test that the dent depth within the CFRP surface at the location of the impact is large enough to be detected during visual inspection. Structural repair is necessary. The challenge for the CFRP design is to find the right balance, i.e. to define a structure thickness that is sufficiently thin (and light-weight), but also tolerant enough against probable impacts to avoid repair effort as much as possible. A very thick CFRP structure would always remain below the detectability threshold so that probable impact energies could not be detected, and thus a repair would never be necessary, but this robustness would burden the aircraft with additional mass, see Fig. 2.18.

The philosophy for CFRP-airframe of "sustain ultimate loads with inner damages if the damage cannot be detected from outside for the complete design service goal of the aircraft without repair" is strongly based on the "no crack growth" assumption of cracks which have been introduced by impact events. This prerequisite limits the maximum acceptable strain level of CFRP. Depending on the material, this maximum strain level is reached at approx. 0.4 %, see Chap. 2. Many fatigue tests have been performed to verify that the residual strength is not negatively impacted as long as these strain limits are not exceeded during cycling. However, the maintenance advantage of CFRP over conventional aluminium

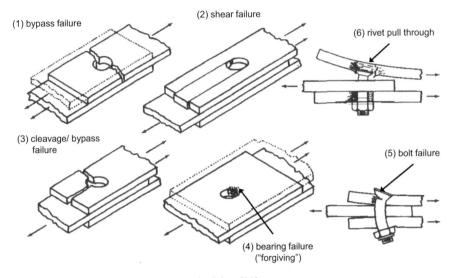

Fig. 6.8 Failure modes of bolted composite joints [11]

structures, which require huge effort during scheduled maintenance in order to detect and assess crack lengths within the aluminium structure, has a certain "price" in terms of wall thickness and weight.

Due to the fact that many probable impact events are relevant for the fuselage structure, sufficient robustness has to be ensured by a minimum skin thickness. With todays toughened 180 °C curing epoxy resin prepregs, a minimum skin thickness of approx. 1.6 mm can be achieved. Except for lightning strike, robustness (demonstrating ultimate load capability for impacts below the detectability threshold) and repair are the main drivers for this minimum thickness. A thickness below 1.6 mm would require improved materials with higher damage tolerance against impact (better compression after impact strength CAI) and improved interlaminar shear strength to avoid rivet head pull through failure of bolted joints.

Rivet head pull through (see Fig. 6.8) is considerable due to the fact that today only bolted repair solutions are certified for structural repair.

Bolted CFRP Repair

Figure 6.8 shows the main failure modes of bolted CFRP joints [11]. For simplification, only one bolt is shown for a double shear joint. The failure modes are as follows:

(1) Bypass failure is not desired. It is probable if the edge distance of the bolt is too low (i.e. the width of the sample related to the diameter of the bolt is too low), or if the portion of 0° layers within the laminate is insufficient.

(2) Shear failure is also not desired and should be avoided. It is probable if the portion of 0° layers in the CFRP laminate is too high.
(3) Cleavage bypass failure is not desired. It is probable if the edge distances of the bolt are too low.
(4) Bearing failure is a sign of a good CFRP design. Compared to all other failure mechanisms, the occurrence of bearing failure (fibre and matrix failure close to the edge of the hole) will not lead to a sudden separation of parts. If the joint is properly designed, the failure stress needed for a total failure caused by bearing loads will be considerably higher than the product of ultimate load and reserve factor. The allowables are usually based on coupon test data for which a 2 % bulging of the hole is accepted, Fig. 6.9. Note that F_u in the force displacement curve of Fig. 6.9 is far above F_{B2}.
(5) Bolt failure is not desired and has to be avoided. It is likely if the bolt diameter is too low.
(6) Rivet head pull through is not desired and should be avoided. It is likely when counter sunk rivets are applied and the minimum cylindrical height within the laminate is too low, and if the ratio of laminate thickness to bolt diameter is too low, Fig. 6.10.

The pull-through effect of counter sunk rivets has also been described in the Composite Materials Handbook [12]. It is evident from Eq. (6.2) that a high material allowable for the interlaminar shear strength ILSS is favourable for thin laminates to avoid undesired rivet head pull through.

$$\tau = \frac{3}{2} * \frac{P}{\pi * D * t} \ [MPa] \tag{6.1}$$

$$RF = \frac{\tau_{ILSS}}{\tau * C_{B-value} * C_{Env} * C_{Bolttype}} \ [1] \tag{6.2}$$

where
   τ is the applied shear load [MPa]
   P is the applied load in bolt direction (see Fig. 6.11) [N]
   D is the diameter of the counter sunk rivet head [mm]

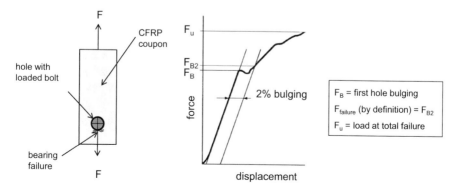

Fig. 6.9 Pin loaded bearing coupon test (*left*) and typical load displacement curve (*right*)

Fig. 6.10 Minimum cylindrical height requirements for countersunk rivets in order to avoid rivet head pull through failure

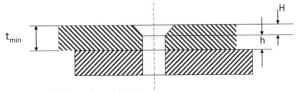

t_{min} = only 1.2 mm for metal structures

t_{min} = 1.6 mm for CFRP structures (i.e. h is > 0.4 mm)!

Fig. 6.11 Rivet head pull through in thin laminates [12]

countersunk side

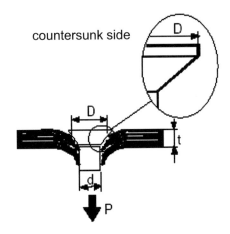

t is the nominal laminate thickness [mm]

RF is the reserve factor (which must be ≥ 1)

τ_{ILSS} is the interlaminar shear strength at room temperature/dry [MPa]

$C_{B\text{-}Value}$ is the B-value factor (to guarantee a high probability of the material allowable) [1]

C_{Env} is the environmental knock down factor (hot/wet condition) [1]

$C_{Bolttype}$ is the bolt factor [1]

The detailed analysis and calculation of bolted joint strength is difficult due to complex stress distribution and failure mechanisms, especially in the area close to the hole edge, and post failure analysis is necessary. Experiments are usually carried out on coupon level.

A possible method can be based on coupon test data gained by pin loaded bearing tests as well as by tension and compression tests with open and filled hole, Fig. 6.12. The vertical axis represents the bearing stress and the horizontal axis the bypass stress of a bolted joint. The green area defines possible bearing/bypass combinations with a reserve factor ≥ 1.

A simple way for the dimensioning of a bolted CFRP joint as depicted in Fig. 6.12, bottom, is to apply the following equations:

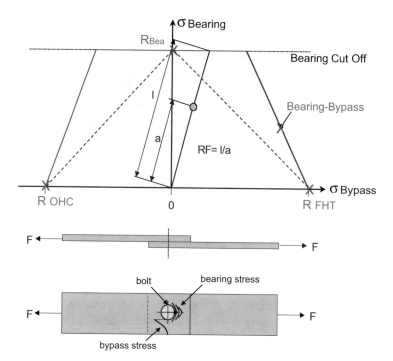

Fig. 6.12 Bearing/bypass graph

$$\sigma_{Bypass} = \frac{F}{(w-d)*t} \ [MPa] \tag{6.3}$$

$$\sigma_{Bearing} = \frac{F}{d*t} \ [MPa] \tag{6.4}$$

where
    t is the nominal CFRP laminate thickness [mm]
    w is the nominal CFRP laminate width [mm]
    d is the diameter of the bolt [mm]
    F is the applied force [N]
    σ is the stress [MPa]
    The graph with the allowable bearing/bypass stress combinations can be defined
by

- experimental determination of the pin loaded bearing strength R_{Bea} (Fig. 6.9; note that the sample width w must be chosen high enough in relation to the bolt diameter d in order to neglect undesired bypass stress influences on the bearing strength), which defines the maximum acceptable stress level ("bearing cut off" line in Fig. 6.12)
- experimental determination of the filled hole tension strength R_{FHT} according to Fig. 6.13 and to Eq. (6.5) (note that the failure force is divided by the gross cross

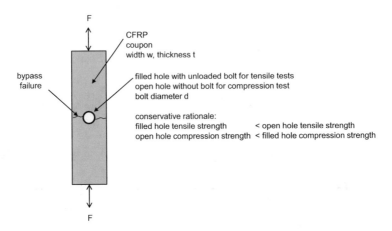

Fig. 6.13 Tension and compression test specimen

section of the CFRP laminate), thus defining the maximum bypass load for tension load cases at zero bearing load, see right side of Fig. 6.12
- experimental determination of the open hole compression strength R_{OHC} Fig. 6.13 and Eq. (6.6) (note that the failure force is divided by the gross cross section), thus defining the maximum bypass load for compression load cases at zero bearing load, see left side of Fig. 6.12

$$R_{FHT} = \frac{F_{failure}}{t*w} \tag{6.5}$$

$$R_{OHC} = \frac{F_{failure}}{t*w} \tag{6.6}$$

where
R_{FHT} is the filled hole tensile strength [N/mm^2]
R_{OHC} is the open hole compression strength [N/mm^2]
w is the nominal CFRP laminate width [mm]
t is the nominal CFRP laminate thickness [mm]
$F_{failure}$ is the failure load [N]

A straight line can be drawn from the position of R_{Bea} to R_{FHT} (for the right "tension side" of the graph) and to R_{OHC} (for the "compression side") in Fig. 6.12. The joint design can be considered as safe if the relevant bearing/bypass stress combination calculated by Eqs. (6.3) and (6.4) is within the triangle created by the dotted line and the vertical and horizontal axis of the graph. However, also combinations with higher bearing stress are possible. Additional experiments with bolted joint coupons can be made to define the slope of the left (compression) and right (tension) boundary line. The reserve factor RF can be calculated by the quotient of 1 to a (maximum allowable stress to applied stress), Fig. 6.12.

Figure 6.14 gives an example of a bolted structural repair of a stiffened CFRP shell, for which several spare parts are needed and the following steps are necessary:

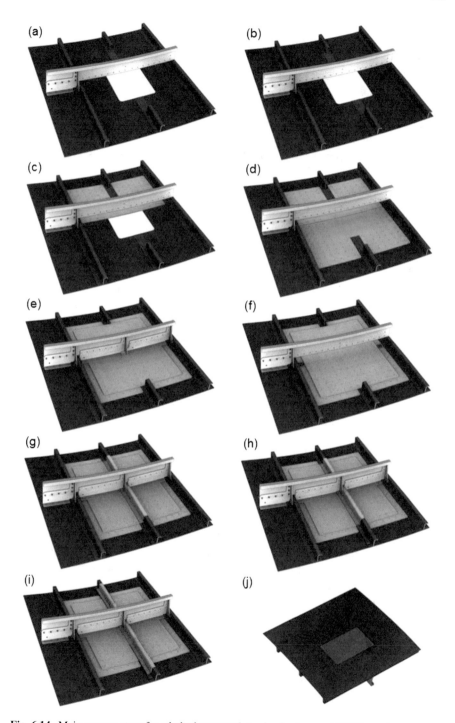

Fig. 6.14 Main process steps for a bolted structural repair of a damaged CFRP panel

(a) the damaged area is mechanically cut out, including the skin and the stringer; in addition, the damaged frame cleats are drilled out
(b) the rivet hole pattern is drilled
(c) the first and (d) second filler on the inside are positioned
(e) an internal doubler and an external flush filler are positioned
(f) fillers are included for the cleats
(g) a new stringer section is included
(h) fillers are included for the stringer couplings
(i) the stringer couplings are positioned and the repair is fully assembled (j)

Bonded CFRP Repair

Although bolted repairs are state of the art and have many advantages, there are also a number of disadvantages:

- significant weight penalty due to the additional material thickness required. Note the strength drop of approx. 50 % from plain tension strength (PLT) to filled hole tension strength (FHT) in Fig. 6.15, and a similar drop of strength properties from plain compression strength (PLC) to open hole compression strength (OHC).
- cost of titanium rivets
- repair effort for drilling and riveting

Adhesive bonding is state of the art for cosmetic repair, but not for structural repair, although it could offer some benefits (see also Chap. 4):

- lower additional weight
- reduced part count
- opportunities for improved architecture and design schemes
- automation potential

However, although a lot of work has been carried out at aircraft manufacturers [13, 14], suppliers [14], operators [15] and R&D institutions [16], the application of bonded structural repair for commercial aircraft is very restricted until today.

Figure 6.16 shows impressive developments of work carried out at Wehrwissenschaftliches Institut für Werk- und Betriebsstoffe WIWeB [16]. A mobile milling system was developed for CFRP repair. Milling of the CFRP in order to prepare the topology for a scarf (chamfered) joint is possible with very high precision, Fig. 6.17. An improved system is described in [17].

Remarkable advancements have also been made in the CAIRE-project, Fig. 6.18, [15], as reported by Lufthansa: *"Specially developed software allows the mobile robot to process 1,000 × 1,000 millimetre surfaces and thick CFRP structures such as wing connection zones. After the damage has been scanned and the surface modelled, the form of the scarfing and the milling path are calculated. The robot grinds out the damaged material, after which the pre-cut fitting of the repair layers is produced. The new part is then manually inserted, glued and cured*

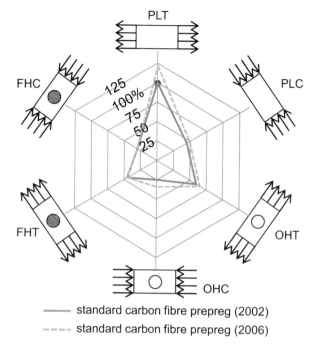

Fig. 6.15 Strength values of a 50/40/10 CFRP laminate of two different prepreg materials [13]

Fig. 6.16 Mobile milling system mounted on a NH90 helicopter structure [16]

in the 3D scarfing surface created by the robot on the fuselage. The new robot enables mobile teams to diagnose and repair large areas of fuselage and wing damage "on wing". By avoiding flights to a maintenance base and by virtue of both shorter ground times and the minimal transport times of individual components, automated repair away from maintenance bases can result in significant cost reductions."

Important developments were also made within the European research project ABITAS "Advanced Bonding Technologies for Aircraft Structures" [14], such as

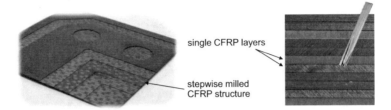

single CFRP layers

stepwise milled
CFRP structure

Fig. 6.17 Precise milling of individual CFRP layers [16]

Fig. 6.18 Damage
scanning and surface
modelling, calculation of
scarfing and automatic
milling [15]. Photography
Gregor Schläger, Lufthansa
Technik

automated surface pre-treatment, development of fast laser triangulation-based fit
analysis systems, development of adhesives, which combine high strength and
durability with flexible process ability such as improved wetting behaviour, multiple curing temperature or a "bonding-on-command" functionality, and monitoring
systems for sensing the physical/chemical state of polymer composite surfaces in an
industrial manufacturing environment, etc.. This is very important, since there are
many sources of surface contamination which can influence the strength of an
adhesive joint very negatively, such as:

- sykdrol
- deicing fluid
- cleaning agents
- dirt
- dust
- oil
- grease
- contaminated water
- hand lotion
- . . .

Figure 6.19 shows lap shear strength values made by IFAM [18], and the
negative influence of surface contamination is clearly visible, see also Chap. 4.

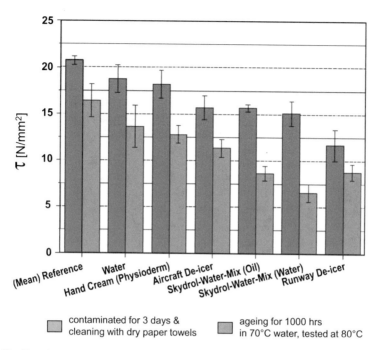

Fig. 6.19 Chamfered CFRP joint lap shear strength (© Fraunhofer IFAM [18])

Among other methods, X-ray fluorescence analysis and FTIR spectroscopy are under investigation to assure CFRP surface quality prior to adhesive bonding. However, the detection of very thin contamination layers and the detection of contamination on mechanically treated, chamfered surfaces are challenging [18].

The main "hurdle" for bonded structural repair is set by the airworthiness authorities. An important source often referred to is CS 23.573 and FAR 23.573 "Damage Tolerance and Fatigue Evaluation of Structure" [19], which is also referred to in [20]:

"The limit load capacity of each bonded joint, the failure of which would result in catastrophic loss of the aeroplane must be substantiated by one of the following methods:

(1) *The maximum disbond of each bonded joint consistent with the capability to withstand the loads [...] must be determined by analysis, tests or both. Disbondings of each bonded joint greater than this must be prevented by design features; or*

(2) *Proof testing must be conducted on each production article that will apply the critical limit design load to each critical bonded joint; or*

(3) *Repeatable and reliable non-destructive inspection techniques must be established which assure the strength of each joint."*

A possible method to fulfil the requirement stated under (1) is to strictly limit the repair size and/or to apply rivets (often referred to as "chicken rivets"). The fulfilment of (2) would mean a very high effort (high cost), as every bonded joint would have to be loaded (tested). Unfortunately, to date there is no non-destructive testing technology available which would fulfil the requirement stated in (3). Although technical solutions have been proposed and filed in a patent [21], these could not yet be successfully demonstrated in practice. Consequently, the applicability of bonded structural repair is very restricted, even though the repair size limits are presently under intensive discussion with the airworthiness authorities [22], and remains an interesting challenge for further R&D work.

Questions

1. When designing CFRP airframe structures, is it necessary to consider repair?
2. What is the advantage of bolted repair? What are the disadvantages?
3. Is bonded CFRP repair possible?
4. How would you ensure that a structural repair of an impact-damaged CFRP airframe structure is not necessary if the damage cannot be detected by visual inspection?
5. How can structural repair solutions be verified?
6. Where are repair solutions and repair methods described?
7. If damage is detected by visual inspection of a CFRP structure, which criteria can be used in order to decide if a repair is mandatory?
8. What does "cosmetic repair" mean?
9. Related to repair, what does "restricted area" mean?
10. What does "temporary repair" mean?
11. Give some examples of "accidental damage".
12. How would you repair a damaged CFRP structure if the residual strength is below ultimate load?
13. Would you expect difficulties for a bolted repair of very thin CFRP skins?
14. Which failure modes of bolted composite joints do you know? When are they more or less likely?
15. Which failure mode of a bolted composite joint would you prefer as a designer? Why?
16. How can the bypass strength of a CFRP laminate be determined?
17. How can the bearing strength of a CFRP laminate be determined?
18. How can allowable bearing/bypass stress combinations of a bolted CFRP joint be determined?

Exercise: Bolted CFRP Repair

1. A load carrying airframe skin structure (material: multiaxial laminate made by the prepreg autoclave technology, 60 % fibre volume content, thickness

1.6 mm) is impacted by a stone. How would you decide if a structural repair is necessary or not?

2. Prepare a graph with the qualitative curve of the residual strength (y-axis) depending on the impact energy (x-axis). Include the ultimate load capability curve. Mark the important repair decision gates.

3. Assume that the stone impact has caused damage, and that the residual strength is below ultimate load. What kind of repair would you prefer, bonded or bolted? Provide some rationale.

4. Prepare a sketch of a simple CFRP test coupon for a single lap joint with two rows of rivets (i.e. two consecutive rivets).

5. What types of failure modes of such bolted CFRP structures can occur?

6. Which failure mode is typical for thin laminates, i.e. with a small ratio of laminate thickness to bolt diameter?

7. Which failure mode is typical for very thick laminates, i.e. with a large ratio of laminate thickness to bolt diameter?

8. Which failure mode would you prefer for a bolted CFRP design? Why?

9. For a simple test coupon of a single lap joint with two consecutive rivets: Assuming that the thickness of the CFRP laminate is t, the width is w, the bolt diameter is d and the force applied at each end of the test coupon is f, and that each of the two bolts will transfer the same load, calculate the bypass tension and the bearing tension. For simplification, assume a linear distribution of stresses.

10. Assuming a ratio of coupon width w to the bolt diameter d of $w/d = 5$, what will be the ratio of bypass stress to bearing stress?

11. Prepare a sketch of a bearing-bypass graph and shade the area of allowable stress combinations.

12. Which kind of coupon test would you propose to assess the maximum bypass tension stress?

13. Which kind of coupon test would you propose to assess the maximum bearing stress?

14. Mark the fictitious test results in the bearing-bypass graph.

15. Mark a fictitious allowable combination of bearing-bypass stress within the bearing-bypass graph. For the ratio of bearing to bypass stress defined by your choice, how could the reserve factor be determined?

References

1. Certification Specifications for Large Aeroplanes (CS-25), Book 2, AMC Subpart D, AMC No.1 to CS-25.603. https://easa.europa.eu/certification-specifications/cs-25-large-aeroplanes
2. Lufthansa Technik. www.lufthansa-technik.com
3. www.aviation-safety-bureau.com
4. Grundlagen der Luftfahrzeugtechnik in Theorie und Praxis, Band II (Flugwerk), Verlag TÜV Rheinland GmbH, Hrsg. Luftfahrt-Bundesamt im Auftrag des BMV, TÜV Rheinland Verlag (1990)

5. International Air Transport Association. http://www.iata.org
6. Wilken, R.: Reparatur von CFK-Primärstrukturen im Luftfahrtbereich. DGLR Conference Proceedings, Stuttgart, Germany, 10.–12.9.2013
7. Hachenberg, D.: Strukturmechanische Anforderungen und Randbedingungen bei der Gestaltung eines CFK-Rumpfes für den Airbus der nächsten Generation, DGLR Jahrbuch 2001, DGLR-2001-133
8. Davis, G.W.; Sakata, I.F.: Design Considerations for Composite Fuselage Structure of Commercial Transport Aircraft. NASA Contractor Report 159296, March 1981
9. SAE International: Chapter 12 – Damage resistance, durability and damage tolerance (formerly MIL-HDBK-17-3F Volume 3). In: Composites Material Handbook CMH-17, vol. 3. SAE International, Warrendale, PA (2012). Published with a Product Code of R-424, ISBN of 978-0-7680-7813-8
10. Fualdes, C.: Airbus – Composite@Airbus, Damage Tolerance Methodology. Presented at the FAA Composite Damage Tolerance & Maintenance Workshop, Chicago, 19–21 July 2006. https://www.niar.wichita.edu/niarworkshops/Workshops/CompositeMaintenanceWorkshop, July2006,Chicago/tabid/99/
11. Hart-Smith, L.H.: Bolted joint analyses for composite structures-current empirical methods and future scientific prospects. In: Kedward, K., et al. (eds.) STP 1455 Joining and Repair of Composite Structures. ASTM, West Conshohocken, PA (2005)
12. Composite Materials Handbook (US Department of Defense). MIL-HDBK-17-1E (issue 2005)
13. Stöven, T.: Rivetless aircraft assembly – a dream or feasible concept. EUCOMAS Proceedings, Berlin, Germany, 7–8 June 2010
14. ABITAS Advanced Bonding Technologies for Aircraft Structures. EU Contract number: AST5-CT-2006-030996, 2006–2009. http://ec.europa.eu/research/transport/projects/items/abitas_en.htm
15. LufthansaTechnik: "CAIRE", BWMI Project 2009–2012, May 2014. http://www.lufthansa-technik.com/innovation-repair-technologies
16. Holtmannspötter, J.: Strukturelles Kleben von CFK Strukturen. Wehrwissenschaftlichen Institut für Werk- und Betriebsstoffe, DGLR Conference Proceedings, Stuttgart, Germany, 10.–12.9.2013
17. Holtmannspötter, J., et al.: Leitkonzept FFS: Das Future Mobile Repair System (FMRS) –Ein Technologiedemonstrator für die automatisierte CFK Reparatur. DGLR Conference Proceedings, Rostock, Germany, 22.–24.9.2015
18. Wolf, M.: Oberflächeninspektion und Oberflächenvorbehandlung für die CFK Reparatur. DGLR Conference Proceedings, Stuttgart, Germany, 10.–12.9.2013
19. FAR/AIM 2015: Federal Aviation Regulations/Aeronautical Information Manual. Skyhorse Publishing, 18.11.2014, Print ISBN 978-1-62914-510-5, Ebook ISBN 978-1-62914-945-5, copyright by Federal Aviation Administration, regulations daily updated on www.faa.gov
20. Advisory Circular AC No 20-107B, U.S. Department of Transportation, Federal Aviation Administration, "Composite Aircraft Structure"
21. Breuer, U., Law, B.: Adhesive joint for joining components of transport craft, in particular of aircraft, and method for determining minimum mechanical load capacity and/or mechanical strength of an adhesive joint. United States Patent US 7,669,467
22. EASA Proposed CM No.: EASA Proposed CM – S – 005 Issue: 01 Bonded Repair Size Limits in accordance with CS 2x.603 and AMC 20-29, 8 Sept 2014

Chapter 7
Flap Design Case Study

Abstract The wing leading edge and trailing edge high lift system is essential for the low speed characteristics of the aircraft. Advanced CFRP technology can contribute to cost and weight savings on aircraft level as well as to improved low speed performance and added product value for the aircraft operator. In this case study, different CFRP flap design schemes and manufacturing technologies are discussed, pointing out the interdependence of CFRP material, design, manufacturing process and resulting part properties. Starting with the description of the reference flap design and the most relevant design targets, new solutions such as multi-spar and sandwich designs are characterised with their advantages and shortfalls. Advanced thermoset solutions (based on prepreg and resin infusion technology) and thermoplastic solutions (based on the tape laying technology) are included. Material and manufacturing as well as strength and stiffness aspects related to aerodynamic performance are analysed. In addition, wing integration challenges linked to new design schemes are discussed.

Keywords High lift • Flap • Torsion box • Galvanic corrosion protection • Erosion protection • Shell sandwich • Multi-spar-design • Braided sleeve • Resin transfer moulding • Thermoplastic tape laying • Recurring cost • Non-recurring cost • Flap gap and overlap deformation • Flap wing integration

General Procedure, Reference Design and Targets

New flap design schemes, materials and manufacturing technologies have been extensively investigated in the course of the R&D program "ProHMS" [1]. The goal was to explore new technical solutions of high lift structure, systems and aerodynamics in order to improve the low speed characteristics of modern aircraft. These characteristics are very important for reducing the noise footprint, allowing steeper approach and descent, improving the take-off performance, enabling landing on short runways, enabling a capacity growth of airports and for many other purposes.

An overview of structural developments which was performed using the A320 as a reference can be seen in Fig. 7.1.

In order to evaluate the cost- and weight-saving potentials of new technologies, detailed studies were performed for the outer flap.

© Springer International Publishing Switzerland 2016
U.P. Breuer, *Commercial Aircraft Composite Technology*,
DOI 10.1007/978-3-319-31918-6_7

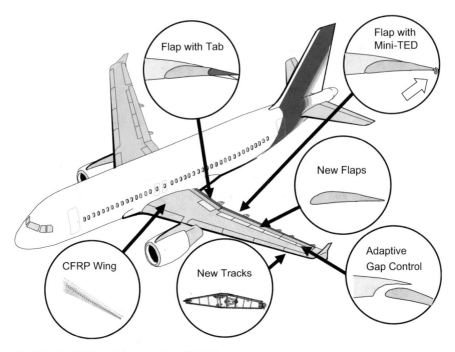

Fig. 7.1 ProHMS studies—overview [1, 2, 4]

The main goals were:

- development of advanced alternative design schemes and efficient manufacturing technologies
- bench marking by comparison of new alternatives with optimised conventional design
- evaluation with assessments of cost, weight, risk and additional criteria
- down selection and proposal of a new design for next generation aircraft

For any new airframe development it is beneficial to define integrated teams with experienced experts from different disciplines, such as:

- aerodynamics
- systems
- loads/weights
- design
- stress
- materials
- manufacturing
- manufacturing tool design
- production
- assembly
- design to cost/value analysis

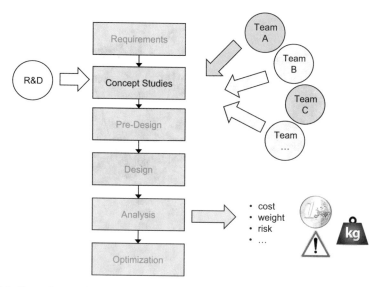

Fig. 7.2 General procedure for design and design evaluation

- maintenance
- other

In the first phase of a completely new development, it can also be beneficial to install multiple teams with the objective of developing different concepts, Fig. 7.2.

Any design effort has to be based on clear requirements (see Chap. 2). In addition, for any cost, weight or risk assessment it is important to have a fully described reference to compare against new solutions.

In ProHMS, the reference for new flap technology was based on the A320 outer flap, Fig. 7.3, consisting of [2]:

- CFRP upper shell
- CFRP lower shell
- CFRP front spar
- CFRP rear spar
- 12 CFRP ribs
- 2 metallic load introduction ribs with metallic fittings
- aluminium-honeycomb-sandwich trailing edge
- CFRP sandwich leading edge
- painting
- joints
- sealings

All CFRP parts are made using the autoclave prepreg technology (see Chap. 4). The torsion box, consisting of front and rear spar, the upper and lower stiffened shell and the ribs, is assembled by bolts and connected to the leading and trailing edge.

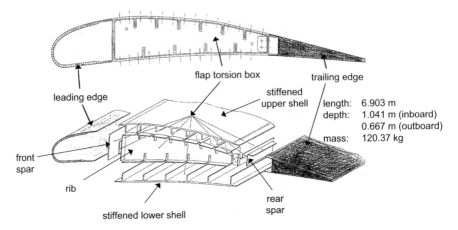

Fig. 7.3 A320 outer flap key characteristics [2]

The following assumptions were made for the design of new concepts (set of preliminary requirements) [3]:

- the recurring cost should be below that of the A320 reference
- the technology should enable a high production rate (high number of flaps produced per month)
- the weight should be below that of the A320 reference
- the outer geometry should be equivalent to the outer A320 flap
- the load introduction points (track positions) should be equivalent to the outer A320 flap
- strength-criteria: maximum skin strain of 0.4 % at ultimate load
- stiffness criteria: deformation limits for gap and overlap as for A320 flaps at 40° (airbrakes retracted) at $v_f = 245$ kts
- stability criteria: no skin and spar buckling up to ultimate load
- aerodynamic smoothness: skin z-deformation <1 mm at $\lambda = 400$ mm for $n_z = 1$, at v_{fe} and at v_c

On top of this, a number of design recommendations were made:

- symmetric and balanced laminates were recommended to avoid warping after curing, and in-plane loading will result in in-plane deformations only
- stacking more than three layers in the same direction (depending on ply thickness) should be avoided
- for load introduction areas a "quasi isotropic" laminate stacking sequence (25/50/25) was recommended
- for CFRP sandwich structures best practice is to use a film adhesive between sandwich face sheets and core
- water ingestion (thin-faced sandwich structures) should be avoided
- for local load introduction areas in sandwich structures inserts were required with local filling of the core

- aluminium-alloys should be insulated from carbon composites in order to avoid galvanic corrosion (typically a GFRP (glass) layer between the Al-alloy and the CFRP laminate provides sufficient insulation), and the aluminium-alloys parts should be specially protected with primer and painted
- Composite panel edges are especially sensitive to erosion when exposed to the air-stream and should be protected with metal; leading edges made of CFRP should be protected with metal

Finally, after the definition of the reference, the most important requirements and design recommendations, it was essential to define and communicate all evaluation criteria and their weightage in order to guide the concept studies according to the overall targets. In the order of decreasing weightage these could be for example:

- recurring cost
- risk
- weight
- operational cost (maintenance effort, repair effort)
- modification friendliness
- non-recurring cost
- other criteria

New Design Concepts

In ProHMS [1], eight different concepts were designed and evaluated, among these

  (i) a shell sandwich design in prepreg-autoclave technology, proposed by a joint team of Airbus and the Corporate Research Centre for Advanced Composite Structures CRC-ACS in Australia
 (ii) a multi-spar design in resin transfer moulding technology, proposed by a joint team of Airbus and Radius Engineering Inc. in the U.S.
(iii) a multi-spar design in thermoplastic tape-laying technology, proposed by a joint team of Institut für Verbundwerkstoffe GmbH in Kaiserslautern, Germany

Shell Sandwich Design

A cross section of the shell sandwich design can be seen in Fig. 7.4.
   The main characteristics were:

- torsion box with sandwich shell (CFRP face sheets and Nomex® honeycomb core) and spar
- two metallic load introduction ribs

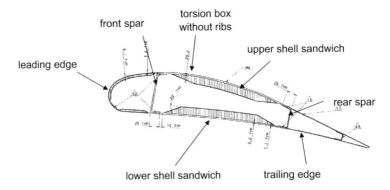

Fig. 7.4 Shell-sandwich design proposed by CRC-ACS [2]

- elimination of all inner CFRP ribs
- prepreg autoclave manufacturing technology
- welded joint of shell and spar
- remaining parts assembled with bolted joints
- trailing edge and leading edge similar to reference

The design and analysis of this concept showed a high weight saving potential compared to the reference of approx. 10 %, a low manufacturing risk due to the application of known materials and technology (except for the welding technology of modified thermoset prepregs; the principle of this technology is explained in Chap. 4, see welding), and a cost reduction potential due to the elimination of ribs [2]. However, the concept could not achieve a fully positive assessment with respect to maintenance and repair, due to the fact that the shell sandwich was judged as more sensitive to impact damages with subsequent water absorption.

Multi Spar Design in RTM

Figure 7.5 gives an impression of the multi-spar design. For the a multi-spar design in resin transfer moulding technology, proposed by a joint team of Airbus and Radius Engineering Inc. in the U.S., the main characteristics were:

- torsion box manufactured in one shot by resin transfer moulding
- elimination of all inner CFRP ribs
- metallic load introduction
- parts assembled with bolted joints
- trailing edge and leading edge similar to reference

The basic process steps for the manufacturing of the multi-spar torsion box by RTM is shown in Fig. 7.6.

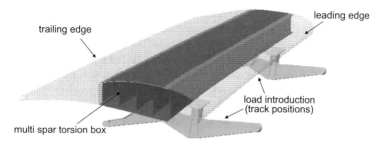

Fig. 7.5 Multi-spar design

Fig. 7.6 RTM multi-spar flap manufacturing process steps: a braided sleeve is pulled over the metal mandrels (1), the metal mandrels are positioned and fixed (2), the setup is positioned within the injection tooling, including additional material for the shells (3), the tool is evacuated, injected with liquid resin and the part is cured under temperature (4). The process has recently been further developed and improved, and it is now also possible to combine textiles, liquid resin and prepreg material in the same tool ("SQRTM", see Chap. 4) [5]. Photography courtesy of Radius Engineering Inc. and Airbus (Bremen)

This design concept showed a very high cost saving potential for non-recurring cost (material cost savings due to the price advantages of textiles and liquid epoxy resin compared to prepreg, and manufacturing cost savings, mainly due to the elimination of auxiliary material linked to the autoclave technology, see Chap. 4) and a high potential for mass production [2]. Although it was proven to be technically feasible to integrate leading and trailing edge in the one-shot multi spar RTM process, it was decided to evaluate a design solution with separate parts due to an easier exchange possibility of leading and trailing edge during maintenance operations. However, detailed investigations were necessary to achieve the weight target for this design, and risks linked to the assembly of load introduction fittings as well as to non-destructive testing had to be examined and assessed in detail.

Since the flap has a tapering aerofoil cross section, decreasing from inboard to outboard, the cross sections between the multi-spars have to follow this taper, and so will the mandrels. This means that the perimeter within each cross section defined by upper and lower shell of the flap and left and right spar will decrease from inboard to outboard. If mandrels with their non-constant cross sections are covered with braided sleeves, the sleeves can easily adapt the mandrel topography, i.e. the complete surface is fully covered by fibres, free of wrinkles or overlaps. However, the mechanism of the cross section and perimeter adaption of the sleeves is fibre shear deformation, Fig. 7.7, see also Chap. 4. If the original braid had a ±45° fibre orientation at a given nominal sleeve perimeter, this orientation would change to ±60° for a larger perimeter, and to ±30° for a smaller perimeter, see Table 7.1. Since both grammage (mass per unit area) and fibre orientation will influence the mechanical properties (strength, stiffness) of the material and the structure, this effect had to be taken into account for a weight optimised design.

Figure 7.8 shows strength properties gained from coupon compression tests with notched laminates made out of braided sleeves by liquid infusion. Starting with the nominal ±45° orientation, the strength value decreases with ±60° orientation. Note that 0° was defined as the load direction. It is obvious that material strength will be

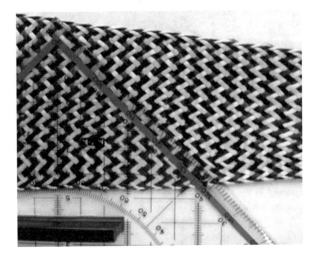

Fig. 7.7 Fibre orientation within the braided sleeve, changing as a function of the sleeve diameter. Note that a carbon fibre—aramid fibre hybrid braid was chosen for this photography only for perceptibility reasons

Table 7.1 Relation of fibre orientation, grammage and perimeter of a braided sleeve

Fibre orientation	±30°	±45° (nominal)	±60°
Grammage	490 g/m^2	425 g/m^2	490 g/m^2
Perimeter	272 mm	385 mm	471 mm

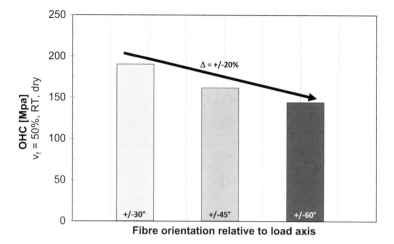

Fig. 7.8 Strength properties (open hole compression strength) of laminates made by braided sleeves with different fibre orientations

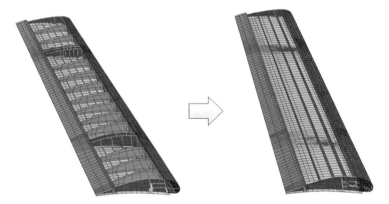

Fig. 7.9 FEM of reference flap (*left*) and multi spar flap (*right*) [1, 3]

negatively influenced if fibres are less oriented in load direction. On the contrary, strength properties are increased for the ±30° orientation.

Stress analysis and design optimisation were carried out by means of FEM, Fig. 7.9 [3]. In addition to the strength criteria, a deformation analysis was also necessary. Figure 7.10, right side, shows a "wash out" effect under load, i.e. the aerofoil deforms in z-direction under the aerodynamic load. Figure 7.10, left side, shows, that this deformation is superposed by a spanwise bending deformation. Note that a superelevation was chosen for perceptibility reasons.

The size of the aerodynamic gap and overlap between wing and extended flap is very important for the low speed characteristics of the aircraft. Figure 7.11 shows a

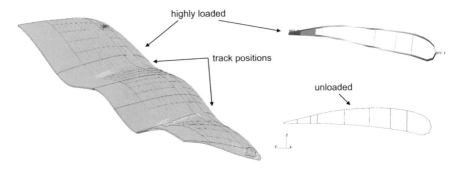

Fig. 7.10 Flap deformation analysis [1, 3]

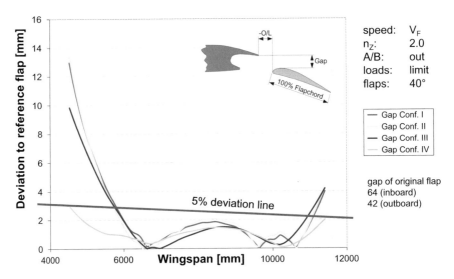

Fig. 7.11 Flap gap deformation analysis for four different multi-spar flap designs [1, 3]

comparative analysis of the deformation behaviour of the reference flap and different design configurations (I–IV) of the multi-spar flap.

It is evident that a stiffness tailoring of the multi-spar flap is not simple as long as no additional means are applied. This is due to the fact that the design freedom of fibre orientation tailoring is limited by the braided sleeves, as their (manufacturing-friendly!) fibre orientation is a function of the part geometry.

It is thus an excellent example for the importance of the understanding and trading of the interdependencies of material properties (weight impact), part geometry (aerodynamic and system installation constraints), design scheme and manufacturing method (cost impact).

A CFRP design is always a "best compromise", and it is decisive for the very beginning of a new design to be clear on the weighing of the evaluation criteria, and the applicable exchange rate between weight and cost. The more light-weight, the more cost, and vice versa.

Multi Spar Design in Thermoplastic Technology

Another interesting design concept was a multi-spar design in thermoplastic tape-laying technology, proposed by the Institut für Verbundwerkstoffe GmbH, Fig. 7.12. The key characteristics were:

- torsion box manufactured with in-situ thermoplastic contour tape laying
- application of consolidated, tailor-made thermoplastic I-profiles as spars
- joining of upper and lower flap shell to I-profiles (spars) by welding during the tape-laying process of the shells
- metallic load introduction
- parts assembled with bolted joints
- trailing edge and leading edge similar to reference

The main benefit of this design and manufacturing proposals compared to the reference technology was manufacturing cost saving (no autoclave process due to in-situ consolidation of the thermoplastic tapes, thus no cost linked to auxiliary materials) [2]. In addition, a high degree of design freedom concerning fibre orientation and stacking sequence of the laminates was evident, Fig. 7.13. However, compared to the multi-spar RTM-concept, the proposal was suffering from disadvantages linked to relatively high material cost for qualified CF/PEEK tape, a lack of manufacturing technology readiness, and from poor surface quality, to name the most important criteria.

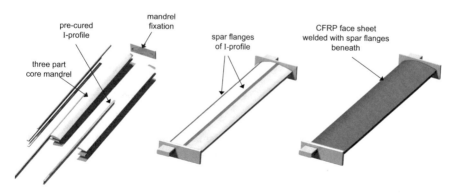

Fig. 7.12 Key process steps of thermoplastic tape-laying of a multi-spar flap (IVW proposal): positioning of consolidated thermoplastic I-profiles (*left*), assembly of mandrels and I-profiles (*middle*), tape-laying of face sheets (*right*) [2]

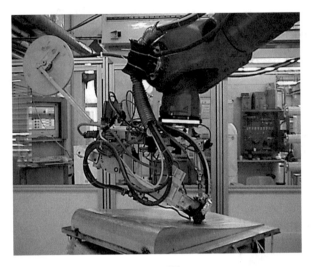

Fig. 7.13 IVW thermoplastic tape laying process [2]

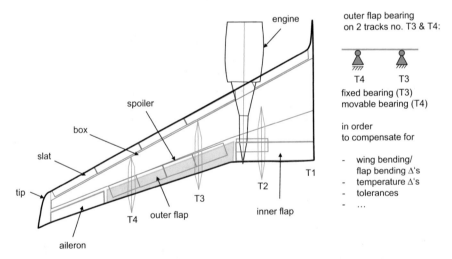

Fig. 7.14 Wing arrangement and position of outer flap (in *blue*), supported by two tracks T3 and T4

For the highly integrated design and manufacturing concepts of the multi-spar flap, interesting questions of accessibility (quality assurance, NDT by ultrasonic testing, maintenance, repair) had to be investigated in detail.

A special challenge was linked to the load introduction points at the track stations, where the resulting air loads are transferred from the flap via the flap tracks into the wing structure.

The position of the outer flap is shown in Fig. 7.14. It is supported by two tracks T3 and T4, where T3 represents a fixed bearing and T4 a movable bearing. The movability in spanwise direction is important to compensate for tolerances of

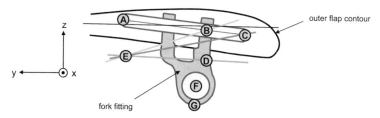

Fig. 7.15 Flap pendulum bearing [1]

different wing and flap bending under load, extensions caused by temperature influences, manufacturing tolerances and assembly requirements.

The principle of the pendulum bearing within the flap torsion box at the track 4 position is shown in Fig. 7.15. The relative movement in x-direction is achieved by the possibility to rotate around the black axis A-B and around the ball-bearing F. The fork fitting connecting the ball-bearing F and the axis A-B is free to tilt due to openings in the lower flap skin. The assembly of the inner pendulum bearing is not a problem for the reference design, since it is a differential design consisting of individual parts for the torsion box.

However, for an integral torsion box design such as the multi-spar design, special solutions are needed. One possibility is to assemble the pendulum bearing A-B outside of the torsion box, i.e. moving the axis to the position E-D. It is obvious from Fig. 7.15 that this is difficult to achieve due to space restrictions, if all adjacent interface parts remain unchanged. In addition, the original lever arm between points F and B is reduced to the much smaller distance of F and D. This leads to a reduction of flap movability in x-direction.

Alternatively, the axis A-C could be tilted to E-C, reducing the x-wise movability less severely, but still requiring an opening of the torsion box for assembly reasons. Another alternative would be to combine an outer pendulum bearing E-D with a repositioned ball bearing G. However, this has an impact on kinematics and on the connection to the track.

Figure 7.16 gives an impression of the space situation between the lower shell of the flap torsion box and the flap track.

This example underlines the difficulties of retrofit solutions, where a clear interface has to be defined and accepted, beyond which further modifications of adjacent structures and systems must be avoided.

Or, to express it positively, it is a good example of the large improvement potential of a fully new designed airframe, where adoptions and optimisation is simpler to achieve.

Fig. 7.16 Space
restrictions between lower
flap shell and flap track

Table 7.2 Evaluation of different flap technologies

Criteria	Shell sandwich	Multi spar RTM	Multi spar TP
Accessibility	0	0	0
Inspection	−	+	+
Maintenance	−	0	+
Repair	−	0	0
Operational cost	−	+	+
Recurring cost	0	+	−
Non recurring cost	+	−	0
Weight	+	−	0
Modification friendliness	+	−	0
Risk	+	0	−

Evaluation

Table 7.2 shows a very simplified evaluation of the different flap technologies. In reality, each criterion must carefully be analysed and quantified in detail by experts of the responsible skill groups, and a full value analysis as well as business case calculations are necessary to prepare a chief engineer decision for the relevant aircraft program.

Questions

1. Name different engineering skills to staff a multi-disciplinary team for the development of new flap design schemes.
2. What is needed to analyse and quantify the potentials of new CFRP design schemes?
3. Name some important criteria to evaluate and quantify new airframe design solutions.
4. Name general recommendations for the design and optimisation of CFRP airframe structures.
5. Are there any specific design precautions for the combination of aluminium and CFRP structures (i.e. a direct attachment)? If yes, why?
6. Except for cost and weight, what are typical design criteria for a CFRP flap?
7. Name some different flap design schemes and describe possible manufacturing routes.
8. Which advantages and disadvantages would you expect from a multi spar design?
9. In case braided sleeves are applied for a multi spar design: In which respect is the design more restricted compared to tape laying technology?
10. Which design relevant CFRP material properties are directly influenced by braided sleeve shear deformation?
11. Comparing different CFRP flap design schemes, would you expect any differences for the flap integration into the wing? If yes, why?

References

1. Prozesskette Hochauftrieb mit multifunktionalen Steuerflächen (ProHMS) BMBF Förderkennzeichen 20F0001C, 1.3.2000–31.12.2002, Zwischenbericht 31.12.2001
2. Breuer, U.P., Siemetzki, M.: Maßgeschneiderte CFK-Strukturen für innovative Hochauftriebshilfen. DGLR Conference Proceedings, DGLR-2002-097, DGLR Jahrbuch 2002, Band I, Deutscher Luft- und Raumfahrtkongress 2002, Stuttgart, 23. bis 26. Sept 2002
3. Bautz, B., Siemetzki, M., Law, B., Latrille, M.: Beitrag zur Entwicklung einer Landeklappe in Vielholmerbauweise. DGLR Conference Proceedings DGLR-2004-028, Deutscher Luft- und Raumfahrtkongress 2004, Dresden, Germany, 20. bis 23. Sept 2004
4. Siemetzki, M., Havar, T.-L., Breuer, U.P.: Structural development of mini-trailing-edge-devices in CFRP. SAMPE Europe Conference Proceedings, Paris, France, 1–3 Apr 2002
5. Gueuning, D., Mathieu, F.: Evolution in composite injection moulding processing for wing control surfaces. SAMPE Conference Proceedings, Sampe Europe 2015, Amiens, France, 15–17 Sept 2015

Chapter 8
Tailored Wing Design and Panel Case Study

Abstract During the development process of a wing it is particularly important to analyse and understand the interaction of external load, resulting internal stresses and component deflection. The latter is strongly influenced by the design. Compared to aluminium, a CFRP wing design offers an additional degree of freedom to tailor the stiffness behaviour for relevant load cases. In this case study, after introducing the most fundamental aero-elastic interrelations and the relevant notions, a historic overview of tailored wing applications is given. The basics of bending-torsion coupling of tailored, non-orthotropic CFRP laminates are explained. Results of aileron efficiency, flutter and lift distribution analysis are discussed, underlining the potential of the application of non-conventional CFRP laminates for wing panels. Aspects of structure strength are analysed by experimental determination of relevant notched laminate properties. Finally, different wing panel design schemes and manufacturing technologies are described, taking into account their applicability for tailored CFRP laminates. This includes skin-stringer integration, tool-laminate interaction, quality aspects such as porosity and distortion, application of functional layers for corrosion and lightning strike protection and other important aspects.

Keywords Aeroelastic tailoring • Wing shape • Wing sweep • Wing torsion box • Isotropic laminate • Orthotropic laminate • Polar diagram • Bending-torsion coupling • Stringer rotation • Laminate rotation • Layer rotation • Aileron efficiency • Flutter • Lift distribution • Drag • Strength properties of tailored CFRP laminates • Wing panel concepts • Differential panel design • Integrated panel design • Tooling concepts • Inner laminate quality • Panel distortion

Wing Shape Control by Aeroelastic Tailoring

Figure 8.1 illustrates the deformation behaviour of a wide body aircraft wing under load. It can be observed that the wing bends in spanwise direction. The green curve in Fig. 8.1 at the leading edge of the wing highlights the bending and the orange line the wing twist.

Where the bending is a simple reaction to the lift forces, the twist can have several route causes, Fig. 8.2. If the wing is modelled by a cantilever beam with a

© Springer International Publishing Switzerland 2016
U.P. Breuer, *Commercial Aircraft Composite Technology*,
DOI 10.1007/978-3-319-31918-6_8

Fig. 8.1 Wing bending (*green curve*) and torsion (*orange line*) during cruise [1]

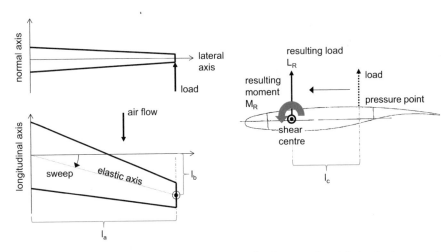

Fig. 8.2 Simplified cantilever model for the wing bending lever arm l_a and the torsion lever arm l_b (*left*). Wing twist caused by the lever arm l_c between pressure point and aerofoil shear centre (*right*)

certain stiffness, a load applied in normal direction will automatically lead to a bending deformation (lever arm l_a). If the wing has a back sweep, as it is usually the case for commercial transport aircraft in order to increase the critical Mach number in transonic conditions, an additional lever arm l_b exists between the upward acting load and the lateral axis indicated in Fig. 8.2, bottom left, leading to a twist θ of the aerofoil. The twist increases from inboard to outboard and reduces the local angle of attack, Fig. 8.3. This means a passive load alleviation: With rising lift forces (for example during manoeuvres), the wing structure will increasingly twist outboard,

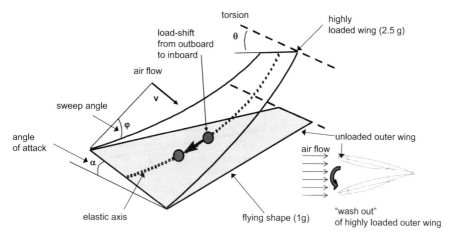

Fig. 8.3 Passive load alleviation of swept-back wing (partly based on [1, 2])

with a so-called "wash out" effect. This decreases the local angle of attack α of the airflow and shifts the resulting pressure point inboard, thus reducing the root bending moment, and the twist moment. The lever l_b arm between the load and the lateral axis is, however, not the only origin of aerofoil twist. Another reason is the lever arm l_c between the chord wise position of the pressure point of the aerofoil, i.e. the point of the resulting lift forces, and the shear centre of the aerofoil structure, leading to a twist moment M_R, Fig. 8.2 right.

A third reason for wing twist can be a structural coupling of bending and torsion by means of a tailored CFRP design. The control of wing torsion has a strong influence on fundamental and overall aircraft performance characteristics. This has been realised very early on by the pioneers of aviation. The fundamental understanding has been strongly supported by scientific work carried out by Shirk, Hertz and Weisshaar [2]. Figure 8.4 provides an overview of wing characteristics and control functions which can be influenced by the torsion behaviour [2].

If—by adequate means—the orientation of the shear centre axis is turned towards the wing leading edge, the "wash out" effect will increase. If the orientation of the shear centre axis is turned towards the wing trailing edge, the "wash out" effect will decrease. The passive load alleviation explained in Fig. 8.3 will not work for a forward-swept wing; on the contrary, this wing will tend to attract more load ("wash in" effect) during manoeuvres, with increasing risk of total structural failure ("divergence"). Divergence protection is possible by a tailored wing design. All other effects (flutter, roll control, etc.) of a tailored, shape controlled wing will be discussed later in this chapter.

Shirk, Hertz and Weisshaar have defined the notion of *Aeroelastic Tailoring* [2]:

Embodyment of directional stiffness into an aircraft structural design to control aeroelastic deformation, static or dynamic, in such fashion as to affect the aerodynamic and structural performance of that aircraft in a beneficial way.

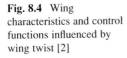
Fig. 8.4 Wing
characteristics and control
functions influenced by
wing twist [2]

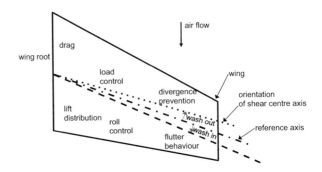

Aeroelastic Tailoring History

Figure 8.5 illustrates Otto Lilienthal, who writes in his corresponding article referring to his attempts to tailor the deflection behaviour of his wings [3]: *"....es war darauf gerechnet worden, dass durch die beim Fluge entstehenden Durchbiegungen die Pfeilhöhe auf ein Zwölftel der Breite, oder noch weiter, herabsinken würde. Die Flügel waren jedoch so steif, dass ihre Krümmung sich nicht veränderte. Es zerschlug sich hierdurch ihre vorteilhafte Anwendung bei stärkeren Winden, die nach den gemachten Erfahrungen eine schlankere Flügelkrümmung benöthigen."*

An interesting early effort to tailor and control the wing wash-in and wash-out behaviour was made in 1935 at Rauanet-Rey in France [2]. The wings were connected to the fuselage by hinge-joints and springs. The point of intersection of these hinges was within the rear part of the fuselage. The system was developed to achieve a wash-out effect (reduction of the angle of attack and inboard shift of loads) during gusts, and a wash in effect for a more efficient roll control during aileron actuation.

In 1954, Short Brothers in Ireland developed the SHERPA S.B.4 (Short & Harland Experimental and Research Prototype Aircraft), a size-reduced prototype of a military long-haul aircraft with aeroelastic tailored wings. The sweep angle of the torsion box was designed to minimise changes of the wing twist during the mission, i.e. to maximise the lift to drag ratio [4–7]. The basic idea was as follows: As the total mass of the aircraft is decreasing during the mission due to its fuel burn, the reacting lifting forces at the wings will decrease, too. Without aeroelastic tailoring, the swept-back of the wings would thus tend to increase their "wash in" during the mission. By means of an aeroelastic tailored design it is possible to minimise unwanted wing twist caused by fuel burn and to maintain an optimised lift distribution (with an optimum ratio of lift to drag) throughout the mission.

In 1969, the U.S. *Air Force Flight Dynamics Laboratory (AFFDL)* initiated a program for the development of aeroelastic tailored wings, leading to numerous intense activities of national airframe manufacturers [2]. General Dynamics developed, manufactured and tested tailored CFRP wings in 1971 and 1974 (3/8 models of F-16 wings, with CFRP wing skins and special UD-layer orientation within the

Fig. 8.5 Otto Lilienthal
1891 [3]

laminates for bending-torsion coupling), including the development of an FEM-based design tool "TSO" = Tailoring and Structural Optimisation. In 1976, the first prototype of a tailored F-16 wing was manufactured [8–13].

In 1977, General Dynamics investigated tailored CFRP-wings for the long-range bomber FB-111 with the target of improving the retention of an optimised lift distribution during the mission by appropriate bending-torsion couplings, thus increasing the range [13].

In 1975, Rockwell developed an unmanned 1:2 prototype of a fighter aircraft with CFRP wings in delta configuration (Highly Manoeuvrable Advanced Technology Aircraft "HIMAT"). The CFRP wing skins had special laminate stacking sequences with defined layer orientations for a bending-torsion coupling. An important design goal was improved roll control [14] (high aileron efficiency).

In 1974, Grumman started aeroelastic tailoring developments [15], leading to a FEM-based tool for optimised CFRP wing structures called "FASTOP" = Flutter and Strength Optimisation Program for Lifting Surface Structures. In 1984, two fighter prototypes "X-29" with forward sweep wings were flight tested, Fig. 8.6. The CFRP wings had anisotropic laminates which were 9° rotated towards the front spar for divergence prevention [17–20]. Long before that, the principles of a swept-forward wing jet aircraft and its advantages such as high angles of attack (late stall and improved stability at high angles of attack as well as low minimum flight speed), high manoeuvrability (due to high aileron efficiency), high lift-to-drag ratio, high range at subsonic speed and short take-off and landing distance had already been studied and extensively flight tested at Junkers in 1944 with two Ju-287 prototypes (1944). Impressive pictures of this aircraft, powered by two forward fuselage mounted Jumo 004-jet engines and four additional wing mounted BMW 003 turbo jet engines can be found in [21]. Engineers of the Ju-287 project (Baade, Wocke) later designed the passenger aircraft HFB320 Hansajet with swept-forward wings (first flight 1964) [22].

Fig. 8.6 Grumman X-29,
NASA's Ames-Dryden
Flight Research Facility,
NASA Photo ID: EC87-
0182-14 [16]

Grumman X-29	
First flight	14.12.1984
Range	560 km
Speed	2019 km/h
Max. take-off weight	7,983 kg

McDonnel Douglas investigated tailored CFRP wings for F-15 and conducted extensive studies for static and dynamic aeroelastic optimisation [23, 24].

Boeing investigated tailored wing designs in 1974 [25], during 1979–1981 tailored CFRP winglets (special UD-layer orientation within the laminates) and aluminium winglets (with stringer sweep) for KC-135 [26].

Gates Learjet Corporation compared aluminium structures with rotated stringer orientation and CFRP structures with rotated UD-layers for swept-forward wings of a passenger aircraft GLC 55 [27].

Lockheed performed studies during 1994–1997 within the frame of the "*PARTI*"-Program (Piezoelectric Aeroelastic Response Tailoring Investigation), funded by NASA, and wind tunnel testing as made with a CFRP wing model which could be deformation-tailored by piezoelectric actuators [28].

Sukhoi developed a fighter aircraft in the 1980s, flight tested in the 1990s, with forward sweep aeroelastic tailored CFRP wing structures, Fig. 8.7 [29].

In Europe, Sensburg at MBB investigated a tailored elongation of the A300 wing, trying to minimise the additional root bending moment [30]. In 1986, tailored CFRP canards were studied for Jäger90 (later: Eurofighter) to increase flutter velocity [31]. At DFVLR the model „TAYLOR" was developed in 1984 to compute bending-torsion-coupling of wings. Between 1997 and 1999, developments were pursued for an aeroelastic tailored outer CFRP wing structure for A380 [1, 32, 33].

Wing Structure Stiffness Tailoring

Figure 8.8 shows a wing with leading edge (slats), trailing edge (flaps, aileron), engine and wing box. The wing box (see Fig. 8.9) is a torsion box, which usually consists of upper and lower panel, spars and ribs.

Fig. 8.7 Sukhoi Su-47 Berkut (Су-47 Беркут—Golden Eagle), originally named S-37 Berkut [29]

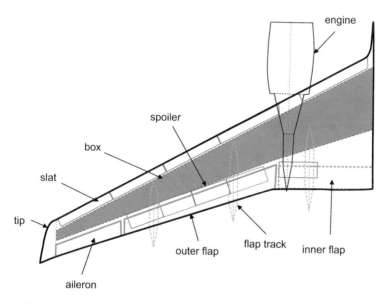

Fig. 8.8 Wing with leading and trailing edge; the wing torsion box is shaded

The wing box panels are usually stiffened with profiles (stringer) to prevent the structure from buckling under load, and, in case of CFRP, the skin usually consists of a multi-axial laminate with unidirectional layers in $0°$, $+45°$, $-45°$ and $90°$. Figure 8.10 shows a generic laminate stacking sequence, in which the same number and mass of $0°$, $90°$ and $-45°$ and $+45°$ layers are used.

A "quasi isotropic" laminate means that the material stiffness properties are independent of the direction in which an in-plane load is applied, Fig. 8.11, case (c). In case different mass percentages of unidirectional layers are used in $0°$, $+45°$, $-45°$ and $90°$ direction, the laminate is anisotropic, and the stiffness changes as a function of the load direction.

The "orthotropic" behaviour of laminates is characterised by the facts that

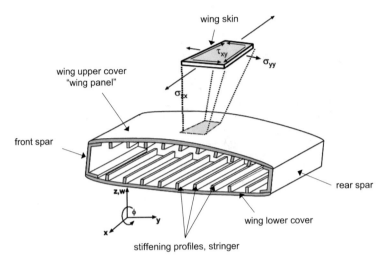

Fig. 8.9 Wing torsion box [34]

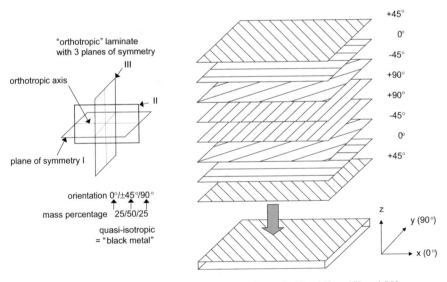

Fig. 8.10 Isotropic and orthotropic laminate with UD-layers in 0°, +45°, −45° and 90°

- the load-deformation behaviour remains unchanged if the material is rotated 180° around one of its three perpendicular orthotropic axis,
- in parallel to the orthotropic axis there is no coupling between normal strain and shear deformation.

This means that in case of an orthotropic laminate any in-plane stress σ_x, σ_y (see Eq. 8.1) is not coupled with shear τ_{xy} and will only lead to ε_x and ε_y deformation,

Fig. 8.11 Typical polar diagram for the laminate stiffness E as a function of load direction for a purely unidirectional fibre orientation with all fibres oriented in 0° direction (a), a bi-axial laminate with equivalent masses of unidirectional fibres in 0° and 90° (b) and a multi-axial isotropic laminate with equivalent masses of unidirectional fibres in 0°, +45°, −45° and 90° (c)

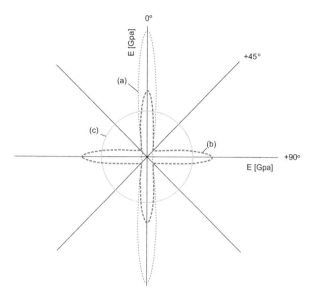

but not to any shear deformation γ_{xy}. For any orthotropic laminates, the terms A_{13}, A_{23}, A_{31}, A_{32} within the stiffness matrix are zero.

This is different for non-orthotropic laminates. Any in-plane stress σ_x, σ_y will automatically lead to a shear deformation γ_{xy}, and vice versa. In this case, the terms A_{13}, A_{23}, A_{31}, A_{32} within the stiffness matrix are different from zero, Eq. (8.1).

This becomes very obvious if a cantilever beam, consisting of a multi-axial laminate, is loaded with a punctual single force positioned at its end within the centre line of the beam, Fig. 8.12.

$$\left\{ \begin{array}{c} \sigma_x \\ \sigma_y \\ \tau_{xy} \end{array} \right\} = \left\{ \begin{array}{ccc} A_{11} & A_{12} & A_{13} \\ A_{21} & A_{22} & A_{23} \\ A_{31} & A_{32} & A_{33} \end{array} \right\} \left\{ \begin{array}{c} \varepsilon_x \\ \varepsilon_y \\ \gamma_{yx} \end{array} \right\} \tag{8.1}$$

In Fig. 8.12, right hand side, due to the shear stress inducing the normal stress (caused by the bending load), the cantilever beam with the non-orthotropic laminate is twisted around its shear centre line. In this generic example, the 0° layers of the laminate have been turned by 10° clockwise. The deformation behaviour can also be understood if we imagine that by turning these layers, more stiffness has been oriented towards the right edge corner of the beam which in consequence is less deflected than the opposite corner.

This stiffness tailoring effect can be used for wing panels. Figure 8.13 illustrates different possibilities. Any rotation of an anisotropic laminate within the wing skin will lead to a different wing twist. This will influence the shear centre line orientation of the wing box to more or less "wash out" or "wash in" deformation, depending on the direction of the rotation, as indicated in Fig. 8.14. Alternatively,

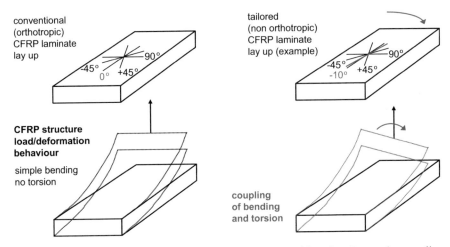

Fig. 8.12 Deformation behaviour of a conventional laminate without bending-torsion coupling (*left*) and a tailored laminate with bending-torsion coupling (*right*). The load is applied within the centre line of the cantilever beam

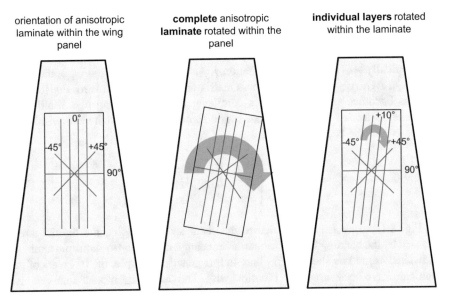

Fig. 8.13 Wing skin (*blue colour*), laminate and layer orientation

or in combination, individual layers within a laminate can be rotated, see Fig. 8.13, right hand side.

Compared to a metal wing box design, for which the orientation of the shear centre line of a given aerofoil and a certain wing sweep can only be influenced by the orientation of the stiffening profiles (stringer), the spars or by a thickness

Fig. 8.14 Shear centre line orientation neutral (**a**), stiffness sweep from the root towards the front spar for wash out (**b**), and stiffness sweep towards the rear spar for wash in (**c**)

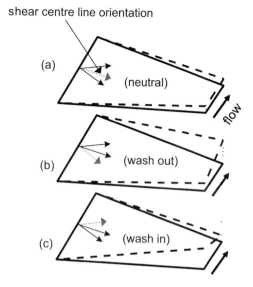

shear centre line orientation

(a) (neutral)

(b) (wash out)

(c) (wash in)

flow

tailoring of the wing skin, CFRP offers additional design parameters due to its laminate structure.

Aileron Efficiency

Ailerons, rudder and elevator are important control surfaces for the manoeuvrability of the aircraft, Fig. 8.15. Ailerons are usually hinged at the trailing edge of the outer wings. An aileron actuation leads to a change of wing lift distribution. More lift is generated if the aileron is moved downwards (the aerofoil camber increases), less lift if it is moved upwards. The change of this lift distribution will cause a moment around the longitudinal axis of the aircraft and it will bank.

In case of elastic swept-back wings, any change of the lift distribution will automatically cause wing twist, Fig. 8.2. The more lift is generated on the wing by the actuation of the aileron (the camber of the aerofoil is increased), the more "wash out" wing twist will occur, Fig. 8.3.

For a given aerofoil, the total resulting lift force is a function of the dynamic pressure q, the wing area A and the lift coefficient, see Chap. 1, Eq. (1.3). The lift coefficient is a function of the angle of attack, Fig. 8.3. The higher the angle of attack, the higher the lift force, and vice versa. The higher the dynamic pressure, the higher the lift, and the higher the resulting wing twist "wash out".

The higher the "wash out" at the outer wing, the lower the local angle of attack, and the lower the lift at this wing position. At a certain dynamic pressure, the wing has twisted to a degree where the aileron is no longer efficient, but on the contrary

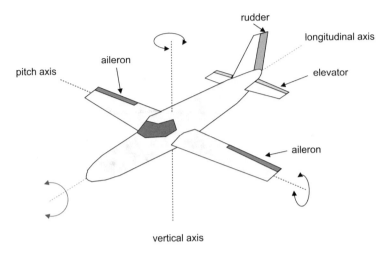

Fig. 8.15 Aircraft control surfaces: aileron (banking, rolling moment, the aircraft rotates around its longitudinal axis), elevator (pitch moment, the aircraft rotates around its pitch axis), rudder (yaw moment, the aircraft rotates around its vertical axis)

can provoke opposite rolling moments. It is advantageous to demonstrate that sufficient aileron effectiveness (and resulting manoeuvrability) is available at very high air speed, the design diving speed v_D, Fig. 8.16, which is far above the cruise speed.

The calculations shown in Fig. 8.16 were made in the early phase of the A380 pre-development, comparing a full metal wing structure to a combination of an inner metal wing with an outer CFRP wing [1]. The black dotted line indicates the aileron efficiency of a conventional flexible metal wing with an ideally rigid outer wing structure as a function of dynamic load. In case of a conventional CFRP outer wing design (anisotropic skin laminate, skin and stringer oriented in parallel to the rear spar), an improvement compared to the full metal wing structure at v_D of approx. 30 % could be demonstrated. The aileron efficiency increased by +50 % for an outer CFRP wing with a different orientation of the anisotropic laminate (0° layers were oriented in parallel to the front spar). A further increase was calculated for the same arrangement, but with a skin stringer orientation parallel to the front spar. The influence of the stringer orientation is remarkable, Fig. 8.17. The aileron efficiency increases approx. 20 % if the stringer orientation is parallel to the front spar instead of parallel to the rear spar.

This can lead to positive snow ball effects. In cases where the size of the aileron is not driven by low speed requirements, the size of high-speed ailerons can be reduced. In case of active load alleviation systems, shifting lift from outboard to inboard by aileron upward actuation, the system effectiveness can be improved, and the maximum root bending moment of the wing structure can be reduced. However, very fast control rates can be necessary for efficient gust load alleviation. In case these control rates cannot be demonstrated, a CFRP wing design tailored to

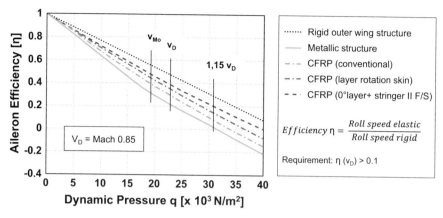

Fig. 8.16 Aileron efficiency [1]. v_{MO} = maximum operating limit speed, v_D = design diving speed

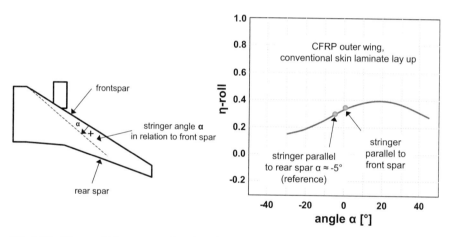

Fig. 8.17 Aileron efficiency η as a function of stringer orientation α

maximise aileron efficiency can lead to unwanted high root bending moments during gusts (due to a poor "wash out" wing twist). Therefore it is necessary to carefully analyse the different load cases.

Flutter

Flutter is an unwanted dynamic instability. The flutter characteristics of a wing are dependent on masses (airframe mass distribution, engine position, etc.), structure damping and stiffness distribution and aerodynamic forces and moments. If the damping of flutter oscillations is negative, amplitude and resulting structure loads will increase. Flutter can lead to the destruction and disintegration of the airframe. For certification it has to be demonstrated that flutter will not occur up to 15 % above the design dive speed v_D (Mach number), i.e. 1.15 v_D/M_D. By means of rotation of anisotropic CFRP laminates or individual layers (see Fig. 8.13) of the wing skin as well as by appropriate stringer orientation of wing panels it is possible to influence resonance frequencies of bending-torsion couplings and to shift the zero crossing points of critical modes to higher velocities.

Lift Distribution and Drag

An important design criterion especially for long haul aircraft is an optimised lift distribution, contributing to minimum drag and minimum fuel burn during cruise. The optimal aerodynamic spanwise lift distribution is elliptic. The local angle of attack—influenced by aeroelastic effects—is very important for the optimal lift distribution. In order to achieve an optimised flying shape of flexible wings, a certain jig shape (shape of the unloaded wing structure) is necessary. However, during cruise, the weight of the aircraft is reduced with its fuel consumption. This reduces the necessary total lift. For elastic swept-back wing structures, a lift reduction means less bending and torsion. Less torsion means that the wing is no longer in its optimal flying shape, and the local angle of attack has changed, Fig. 8.3. If, for example, the lift distribution and the resulting wing structure is optimised for the mid of a certain long haul mission for a given constant altitude and at a given constant speed, a conventional elastic wing structure will tend to a "wash out" twist during the beginning of the mission. In order to obtain the necessary total lift, the aircraft has to be operated at a higher angle of attack, leading to a reduction of the lift to drag ratio. During the end of the mission, due to fuel burn, the twist will be reduced, and again the lift distribution will be not optimal [1]. In order to maintain an optimal lift distribution, the flying shape should remain unaffected by the mass reduction. This is possible by tailoring the bending-torsion behaviour, Fig. 8.18. The difference between jig shape and flying shape $\Delta\alpha$ has exemplarily been calculated for the maximum take-off weight (MTOW) and the maximum zero fuel weight (MZFW) of a generic aircraft. In case of a conventional CFRP wing design (anisotropic skin laminate, skin and stringer oriented in parallel to the rear spar), $\Delta\alpha$ is approx. 1°. This delta can be significantly reduced for a tailored CFRP design (anisotropic skin laminate, skin laminate and stringer oriented in parallel to the front spar). On the contrary, $\Delta\alpha$ becomes larger in case of an adverse tailoring, with a laminate and stringer orientation parallel to the rear spar. For A340, a $\Delta\alpha$ reduction of only 0.5° leads to an improvement of the lift to drag ration L/D of approx. 1 %, meaning 400,000 l less fuel burn per aircraft and year [1].

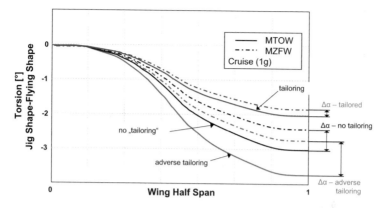

Fig. 8.18 Wing twist (delta of jig shape—flying shape) as a function of wing half span and tailoring

Strength Properties

Since the wing structure design is not only driven by its aeroelastic behaviour but also by stability and strength requirements, it is necessary to take the relevant material properties into account. Figure 8.19 shows important strength properties of anisotropic notched CFRP coupons as a function of the off-axis-angle in which the specimen were loaded. The tailoring principle is the one shown in Fig. 8.13 (mid). For open hole compression (OHC) and open hole tension (OHT) a clear decrease of strength is obvious, as the majority of load carrying 0° fibres is turned away from the main load direction. For the pin loaded bearing (PLB) strength, however, the drop is less severe, and for the ultimate PLB strength an increase can be observed. This is due to the well-known phenomenon that the ultimate bearing strength is supported by fibres oriented with an inclination angle relative to the main load direction. A similar behaviour can be observed for tailored anisotropic laminates with rotated 0° fibres, Fig. 8.20. The tailoring principle is the one shown in Fig. 8.13 (right). In case laminates with rotated 0° fibres are used, their off-axis strength properties must be analysed as well. For the OHC and OHT coupon tests it is important to note that additional shear is applied if the load direction deviates from the main stiffness orientation of the anisotropic laminates. The shear deformation of the coupons, however, can be inhibited by fixed sample clamping. It should be ruled out by a sufficient free sample length in between the load introduction clamping that the influence of additional shear load caused by the fixed clamping on the failure (i.e. on the critical stress distribution around the notch in the mid of the sample) is negligible. This has been verified for the tests performed according to AITM in [1].

When tailored laminates are applied for wing skins according to the principles shown in Fig. 8.13 mid or right, or for the combination of these principles, a stress analysis with a space-resolved view of main stress directions and main laminate stiffness orientation will reveal the fact that the off-axis load angles will be less severe (i.e. smaller) than expected. This is due to the fact that the load within the structure will tend to follow the main stiffness paths.

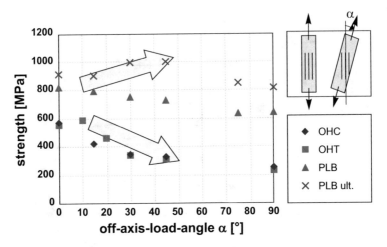

Fig. 8.19 Laminate strength properties under 0° and off-axis load angle α. Prepreg with standard carbon fibres, the cured ply thickness (cpt) was 250 μm, the fibre volume content v_f 60 %, the laminate stacking sequence $(+45,0_2,-45,90,+45,0_2,-45,0)_s$ 50/40/10, the hole diameter was 6.35 mm, the sample thickness was 5 mm, tests were performed according to AITM [1]. Only mean values of six samples per measurement are shown, the standard deviation was 5–10 %

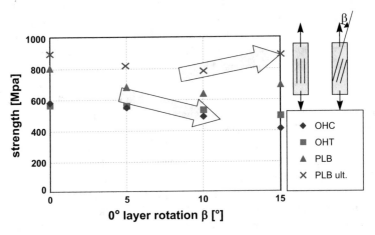

Fig. 8.20 Laminate strength properties as a function of 0° layer rotation angle β. Prepreg with standard carbon fibres, the cured ply thickness (cpt) was 250 μm, the fibre volume content v_f 60 %, the laminate stacking sequence $(+45,0_2,-45,90,+45,0_2,-45,0)_s$ 50/40/10, the hole diameter was 6.35 mm, the sample thickness was 5 mm, tests were performed according to AITM [1]. Only mean values of six samples per measurement are shown, the standard deviation was 5–10 %

Wing Panel Concepts

An integral wing skin design and manufacturing concept which has already been applied for VTP shells [1] is shown in Fig. 8.21. Prepreg material is placed on split aluminium mandrels. This CFRP material will later form the stringer and rib

Fig. 8.21 Integral "wet in wet" technology with I-shaped stringer and aluminium split mandrels, integration by co-curing

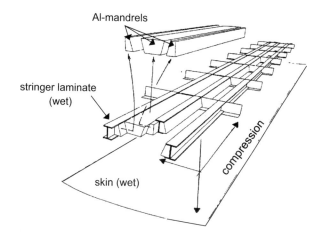

attachment brackets. The metal mandrels are positioned and mounted on a rig and mechanically pre-compacted. The complete set-up is placed on top of a wet skin laminate (prepared by tape laying) and subsequently vacuum bagged. During the following autoclave cycle, skin and stringer prepreg is material cured, i.e. skin and stringers are co-cured. The final compaction of the stringer blades is achieved by means of the thermal expansion of the aluminium mandrels. The volume of the stringer blades is defined by the contour of the aluminium mandrels, meaning an isochoric process in these areas.

Compared to alternative concepts discussed in this chapter, this technology has the following advantages:

- One-shot manufacturing of panels, i.e. only one autoclave cycle for skins and stringer is necessary
- Integration of rib attachment brackets
- Span wise adjustment of stringer cross section to force flow possible (weight optimisation)
- No stringer couplings needed (less rivets, less mass, less assembly effort)
- Large degree of design freedom for stringer orientation

On the other hand, compared to alternative concepts, there are also disadvantages:

- A large number of individual, expensive metal mandrels are needed
- A fully automated application of prepregs to mandrels can be difficult if the prepreg "tack"(its stickiness) is non-constant
- High manual effort for mandrel demoulding
- High effort for mould cleaning
- High invest for compaction and mandrel placement machinery (mechanical pre-compaction transversal to stringer orientation) as well as for the transfer and turning machinery
- Low flexibility for thickness adaptations, i.e. design changes mean new mandrels

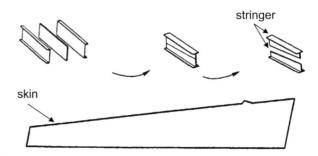

Fig. 8.22 Differential technology, stringer or skin are made separately and integrated in a second step by co-bonding

- Local variations in prepreg matrix content can cause quality issues (porosity) due to isochoric consolidation in the stringer blade
- Sharp edge in stringer feet—skin transition area
- High effort for glass ply integration inside the shell (insulation layer for corrosion protection or for other purposes)

An alternative concept is shown in Fig. 8.22. The stringer profiles are manufactured separately. In case double T-stringer profiles are manufactured, they can be cut lengthwise to obtain T-stringer profiles. Cured stringer profiles are placed on a wet skin (prepared by tape laying) and bagged for a subsequent autoclave cycle. The stringers are attached to the skin during the curing of the skin by co-bonding. It is common practise to apply additional adhesive film material between skin and stringer, as this can enhance fracture toughness and crack opening energy (G_{ic}). If the stringer profiles are manufactured with a constant cross section (by pultrusion, for example, see Chap. 4), different stringer families with different cross sections can be placed on the skin: A first family with the largest cross section inboard followed by a second family with a smaller cross section etc., i.e. reducing the cross section with the force flow. However, this requires mechanical stringer couplings, and the additional weight for these couplings must be traded against the weight advantage of reduced stringer cross sections. Alternatively it is also possible to use individually weight optimised stringer profiles with tailored tapering cross sections, however, compared to constant cross section profiles, this increases the stringer manufacturing cost.

The following advantages can be summarised:

- Subcontracting of stringer manufacturing and separate quality check of stringer profiles is easily possible
- Standardisation by definition of stringer families (with stringer profiles of the same cross section)
- Relatively low tooling effort for panel manufacturing
- Relatively low rigging and demoulding effort

However, there are also disadvantages:

- Stringer couplings are necessary in case of a stringer family concept, causing additional assembly effort and additional mass

- Maintaining the exact stringer position is difficult on a wet skin; low matrix viscosity at higher temperatures and unwanted external forces on stringer blades caused by the vacuum bagging can result in unwanted position changes
- Assembly difficulties can occur as a result of stringer position tolerances
- Separate assembly process for rib attachment brackets
- Skin layer displacements within the transition area of hard stringer feet to wet skin can occur; expensive auxiliary rubber material (pressure pads) is needed to ensure a more constant pressure distribution in these areas during consolidation
- Sharp edge in stringer feet—skin transition area

Another alternative concept is shown in Fig. 8.23. Prepreg material is processed to planar stringer laminates (2D-tape laying, see Chap. 4). After cutting, stripes are vacuum-hot-formed to U-shaped profiles by means of infrared radiation and flexible diaphragms above metal mandrels. Mandrels with U-shaped laminates are placed on a wet skin (prepared by tape laying) and vacuum bagged. During the subsequent autoclave process, stringer and skin are co-cured. The thermal expansion of the metal mandrels can be used to apply consolidation pressure on the stringer blades.

This concept has the following advantages:

- One-shot process, only one autoclave cycle
- Less metal mandrels compared to the integral concept shown in Fig. 8.21
- Precise stringer positioning
- No stringer couplings needed (less rivets, less mass, less assembly effort)
- Simple application of insulating glass ply on inner panel side
- No stringer—skin edge

In contrast, the following disadvantages must be considered:

- Limited degree of design freedom with respect to stacking sequences of stringer blade and skin, since the upper skin layers are at the same time part of the stringer blades
- Design changes of stringer blade thickness require new tooling
- Local variations in prepreg matrix content can cause quality issues (porosity) due to isochoric consolidation
- Separate assembly process for rib attachment brackets
- The stringer blade position must be perpendicular to the tool plane in order to avoid demoulding difficulties

Fig. 8.23 Integral design with wet U-shaped stringer profiles on a wet skin, integration by co-curing

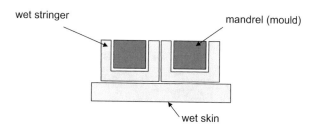

- High invest for compaction machinery (mechanical pre-compaction transversal to stringer orientation) and for transfer/turning machinery

Another possibility is shown in Fig. 8.24. Skins are separately manufactured and cured. Stringer profiles are manufactured by tape placement and cutting of planar laminates, and by subsequent vacuum-hot-forming of these prepreg-stripes on L-shaped moulds. These moulds are mechanically milled metal parts with a thickness of typically more than 10 mm. A stringer profile is made by adverse positioning of L-shaped moulds (vis-à-vis) and stringer laminates. Moulds remain in position during autoclave stringer cure and stringer co-bonding to the pre-cured skin.

This concept has the following advantages:

- High flexibility for thickness variations of stringer blades, stringer feet and skin; L-shaped moulds can be used for different stringer thicknesses, only filler material for radii needs to be adapted
- No stringer couplings needed (less rivets, less mass, less assembly effort)
- Less tooling effort compared to separate manufacturing of individual stringer profiles
- No negative impact of stringer on skin laminate, i.e. no negative effects of local pressure variations on the skin within the skin—stringer transition area
- Less auxiliary material needed (silicone rubber pressure pads etc.) during stringer/skin manufacturing
- Simple autoclave set-up and manufacturing of skins
- Simple quality check (fast C-scan) of pre-cured skins
- Precise positioning of stringers can be achieved by local fixation of stringer tooling and skin tooling
- Local thickness variations/tolerances of the skin can be equilibrated by wet (compressible) stringer lay up

However, the following disadvantages must also be considered:

- Two autoclave cycles are necessary
- High tooling effort to handle long L-shaped stringer moulds
- Relatively high effort for application of insulating glass plies on the inner surface of the stiffened panels
- Sharp edge in stringer feet—skin transition area

Fig. 8.24 Differential technology with wet stringer on a hard (pre-cured) skin, integration by co-bonding

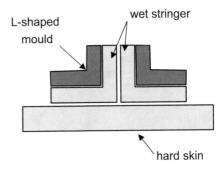

- Separate assembly process for rib attachment brackets

A last example is shown in Fig. 8.25, [35]. Prepreg material is processed to planar stringer laminates (2D-tape laying). After cutting, the prepreg stripes are vacuum-hot-formed to U-shaped profiles by means of infrared radiation and flexible diaphragms above metal mandrels. Mandrels with U-shaped laminates are placed on a wet skin (prepared by tape laying). A pre-cured stringer blade is positioned in between adjacent sides of the U-shaped laminates. All mandrels are removed, and supporting tools are placed to support the stringer profiles during the subsequent autoclave process. Stringer and skin are co-cured. The individual process steps are highlighted in Fig. 8.26.

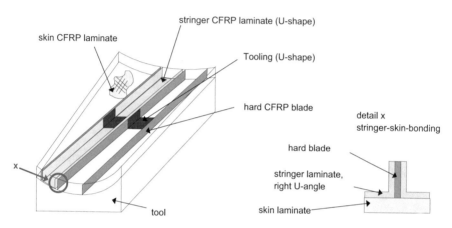

Fig. 8.25 Integral technology with hard (pre-cured) stringer blades and wet u-shaped stringer/skin laminates on a wet skin, integration by co-curing (skin) and co-bonding (stringer), "hard in wet on wet" concept [35]

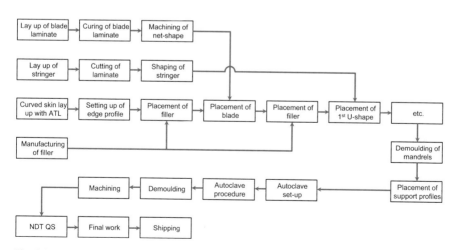

Fig. 8.26 Basic process steps for stiffened shell manufacturing ("hard in wet on wet" concept)

This "hard in wet on wet" concept has the following advantages:

- One-shot process, only one autoclave cycle is needed
- No stringer couplings are needed (less rivets, less mass, less assembly effort)
- Less metal mandrels compared to the concept shown in Fig. 8.21
- Simple application of insulating glass ply on inner panel side
- No sharp stringer—skin edge
- Consolidation at isobaric conditions, lower risk of quality issues (porosity) in case of locally varying matrix content in the prepreg material
- Stringer cross section adaption to force flow (weight optimised design) possible through tapered blades
- Flexibility for late design changes (thickness changes), but tooling for stringer shaping must be adapted

On the other hand, the following disadvantages are obvious:

- Limited degree of design freedom with respect to stacking sequences of elements, since the upper skin layers are at the same time part of the stringer blades
- The stringer blade position must be perpendicular to the tool plane in order to avoid demoulding difficulties of the supporting tools
- Separate assembly process for rib attachment brackets
- Difficult handling of long metal mandrels
- Difficult positioning of wet U-shaped laminates on skin laminate
- Difficult integration of radii filler materials

Special Manufacturing Aspects

When new wing skin concepts are evaluated and compared for a new aircraft program, not only weight and cost assessments, but also risk assessments are necessary. All phases have to be taken into account (manufacturing, assembly, handling, transport, test, operation, maintenance, recycling). For the skin stringer integration process (manufacturing phase), among other aspects it is particularly important to investigate

- hard-wet interface bonding quality
- laminate compression and inner laminate quality (porosity)
- stringer positioning
- doubler integration
- lightning strike protection application (usually an expanded copper or bronze foil of 70–200 g/m^2)
- application of corrosion protection, especially at interfaces to aluminium parts to avoid galvanic corrosion (usually by means of glass fibre prepreg)
- panel distortion

Figure 8.27 illustrates the compression and consolidation of a stringer stiffened shell according to the "hard in wet on wet" concept during the autoclave process

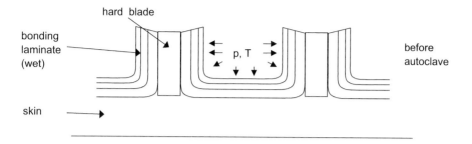

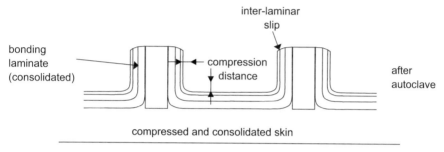

Fig. 8.27 Compression and consolidation of a stringer stiffened shell

Fig. 8.28 Inter-laminar slip of a u-shaped laminate (see arrow)

under temperature and pressure [35]. In order to obtain the needed inner laminate quality free of voids and the final part thickness, it is necessary to enable interlaminar slip, Fig. 8.28.

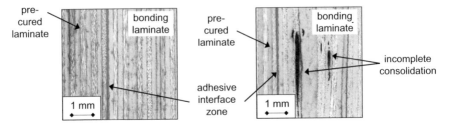

Fig. 8.29 Co-bonding process of pre-cured and wet laminates. Optimal inner laminate quality (*left*) and poor laminate quality with voids (*right*)

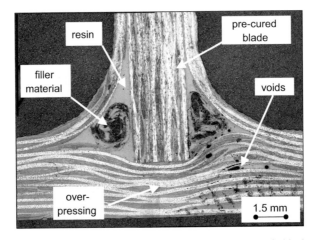

Fig. 8.30 Stringer foot–skin interface with significant over pressing of skin layers under the stringer blade and adjacent poor laminate quality with voids

In case supporting tools are used to prevent the stringer blades from unwanted tilting, Fig. 8.33 (bottom right), it is important to investigate the laminate quality under these tools. It has to be insured that sufficient pressure can be applied to the CFRP underneath these supporting tools. Figure 8.29 illustrates sufficient laminate quality (left) and poor quality with voids (right). Sufficient pressure can be applied with aluminium tools and their thermal expansion.

Another important area is the stringer blade—skin interface. Special filler material can be necessary in the radii, Fig. 8.30. For co-bonding processes, in which hard (pre-cured) elements are positioned on a wet lay-up, it can be necessary to apply additional rubber pads in order to obtain a homogeneous pressure distribution. The unwanted warping and insufficient consolidation of layers with voids shown in Fig. 8.30 is typical for uneven pressure distribution. In this case, where a pre-cured stringer blade was positioned between wet U-shaped stringer and on a wet skin, the pneumatic autoclave pressure on the vacuum bag was unevenly distributed. A telegraphing effect of the vacuum bag concentrated pressure on the stringer blades, leading to an over pressing of the wing skin and to a lack of pressure

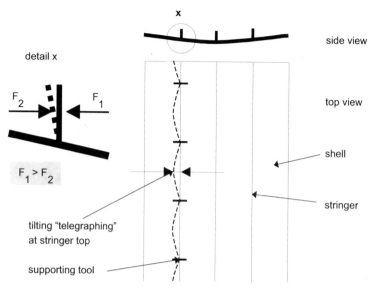

Fig. 8.31 Skin stringer arrangement with discrete stringer fixation and telegraphing effect

in the adjacent areas. These problems can be avoided with triangular rubber pads, distributing the autoclave pressure more evenly.

A different type of a telegraphing effect is shown in Fig. 8.31. In case of a discrete stringer blade fixation during the autoclave process, it is possible that the stringer blades tend to tilt if—as a result of geometry—the force generated by the autoclave pressure and the area of the left stringer side is different from the opposite side.

Distortion can be a problem especially for very thick and stiff shells. Residual stresses during assembly should be avoided. Adjustment of distorted parts by shimming the interfaces is a time consuming and expensive process. Figure 8.32 illustrates a "spring in" distortion of a stringer stiffened shell.

Thermal expansion coefficients differ in longitudinal (fibre-axis, $\alpha_{longitudinal}$) and transversal (axis perpendicular to fibres, $\alpha_{transversal}$) direction, Table 8.1.

With reference to typical coefficient of thermal expansion values given in Table 8.1, a very simplified thought experiment can help to explain the formation of the distortion during the autoclave curing process:

- the temperature rises from room temperature to cure temperature (180°)
- the material is still "soft"
- there are no internal stresses yet
- the material cures at constant temperature (180°)
- the matrix status changes from soft to hard
- no internal stresses occur if the shrinkage of the matrix is neglected
- the material cools down to room temperature
- the carbon fibres try to extend in their longitudinal direction
- the matrix tries to shrink in all directions

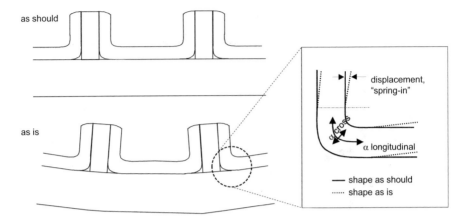

Fig. 8.32 Spring in of a stiffened CFRP shell

Table 8.1 Approximate
thermal coefficient values for
carbon fibres (CF), epoxy
resin and CFRP (60 % fibre
volume content)

Material	Thermal coefficient α [μm/m K]
CF_{trans}	+5
CF_{long}	−0.7
Epoxy	+55
CFRP UD_{long}	+0
CFRP UD_{trans}	+35

- the fibres try to shrink perpendicular to their main axis
- internal stresses occur
- the stresses are partly released through spring-in deformation

Further aspects have to be taken into account: During the curing process and the applied pressure, the laminate is consolidated and compacted, meaning that its thickness decreases. Usually, but depending on the part thickness and pre-compaction processes, the cured laminate thickness is about −10 % compared to its non-cured wet thickness for standard prepreg material. In order to form radii an interlaminar shear (interlaminar gliding, slipping of single prepreg layers) is necessary, see Figs. 8.27 and 8.28. Note that during the hot pre-forming process of the U-shaped bonding laminates, interlaminar slip has already occurred, but additional slip is necessary during compaction to the final cured laminate thickness. If, however, interlaminar slip is inhibited (due to high matrix increasing viscosity, friction forces, . . .), the compaction of the laminate in its flat areas leads to in-plane forces and internal stresses in the outer radii layer, i.e. the outer radii layers are under tension. After final cure and de-moulding, these tension stresses can partially be released through the spring-in deformation of the shell.

In addition, when simple L-shaped elements (tools) are used to support the stringer blade positioning and fixation (see Fig. 8.33 bottom right), unwanted tilting or deformation of these angles (L-tools) can occur during the application of pressure within the autoclave. If the counter-acting compaction resistance of the

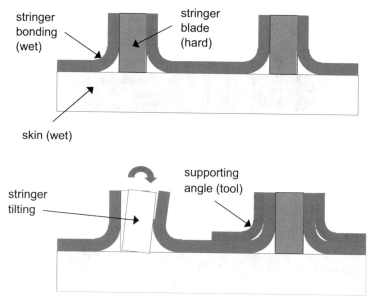

Fig. 8.33 Idealised stringer arrangement (*top*) and unwanted stringer tilting (*bottom*)

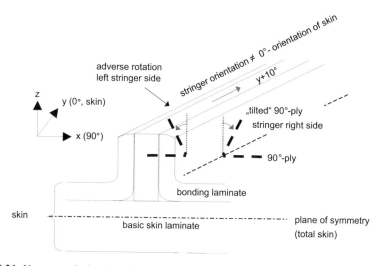

Fig. 8.34 Unsymmetrical stringer bar laminate due to stringer rotation

laminate under the L-profiles is different for the skin laminate and the stringer blade laminate, the L-profiles left and right of the stringer blade can rotate, i.e. the left L-profile clockwise, and the right L-profile anticlockwise, creating curvature in the skin laminate, Fig. 8.32 bottom, thus supporting the spring-in-effect.

When comparing different skin—stringer integration technologies it is also important to investigate effects of the stringer orientation with respect to the orientation of the 0° layer in the wing skin, Fig. 8.34. In case the stringer blades

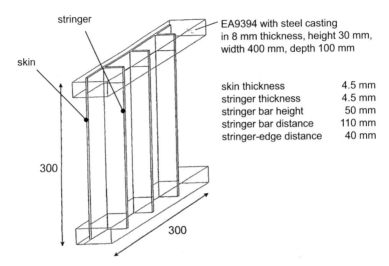

EA9394 with steel casting
in 8 mm thickness, height 30 mm,
width 400 mm, depth 100 mm

skin thickness	4.5 mm
stringer thickness	4.5 mm
stringer bar height	50 mm
stringer bar distance	110 mm
stringer-edge distance	40 mm

Fig. 8.35 Stiffened CFRP compression panel

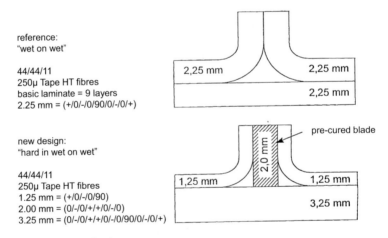

reference:
"wet on wet"

44/44/11
250μ Tape HT fibres
basic laminate = 9 layers
2.25 mm = (+/0/-/0/90/0/-/0/+)

new design:
"hard in wet on wet"

44/44/11
250μ Tape HT fibres
1.25 mm = (+/0/-/0/90)
2.00 mm = (0/-/0/+/+/0/-/0)
3.25 mm = (0/-/0/+/+/0/-/0/90/0/-/0/+)

Fig. 8.36 Skin–stringer laminates

or part of the stringer blades are built with the upper layers of the skin, any deviation of the stringer orientation from the 0°-layer of the skin will lead to unsymmetrical laminates of the stringer blades.

In general, the different interacting effects leading to distortion are very complex, and in some cases it can be necessary to adapt the tool geometry in order to obtain the needed part shape.

Figure 8.35 shows a compression test panel which can be used to determine the compression strength of different skin—stringer integration technologies. In this example, the influence of an integrated pre-cured stringer blade was assessed by comparing the two different integration technologies shown in Fig. 8.36.

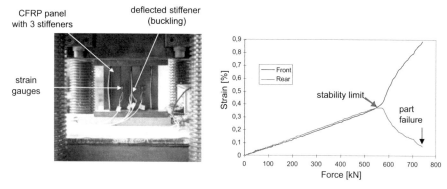

Fig. 8.37 CFRP compression panel test set up and "chute"-*curve* of measured strain (front and rear mounted strain gages)

Table 8.2 CFRP panel compression test results

	"Wet on wet" design				"Hard in wet on wet" design			
Compression shell no.	1	2	3	Average	1	2	3	Average
Stability limit	520 kN	580 kN	565 kN	555 kN	500 kN	500 kN	500 kN	500 kN
	3.35 ‰	3.65 ‰	3.70 ‰	3.56 ‰	3.30 ‰	3.30 ‰	3.35 ‰	3.32 ‰
Strength	721 kN	736 kN	742 kN	733 kN	667 kN	594 kN	612 kN	624 kN
	8.00 ‰	8.20 ‰	8.25 ‰	8.15 ‰	7.40 ‰	5.90 ‰	6.80 ‰	6.70 ‰

The test set up and a test result is shown in Fig. 8.37. When the panel is compression loaded, the stability limit can be determined by the occurrence of buckling. In this case, strain gauges mounted on the front and back side of the panel indicated the stability limit: during buckling, compression forces increase on one side of the panel and decrease on the opposite side ("chute"-curve).

The compression test results are shown in Table 8.2. Both the stability limit and the average compression strength decrease in case of the integration of a pre-cured blade, but the drop in compression strength is more severe. Where the load transfer between stringer blade and skin is relatively smooth in case of the "wet on wet" reference, Fig. 8.36, the layers are interrupted at the bottom of the hard blade, leading to lower compression strength.

Table 8.3 summarises some important features of the different skin-stringer integration technologies.

Table 8.3 CFRP skin–stringer integration technology comparison

Feature	Integral "wet in wet" I stringer and split mandrels	Integral "wet in wet" U-shaped straight	Differential "hard on wet"	Differential "wet on hard"	Integral "hard in wet on wet"
Stringer bar angle related to tool plane	Any	Perpendicularly to tool plane	Any	Any	Perpendicularly to tool plane
Stringer rotation/ curved stringer	Any	Any	Limited (for straight profiles)	any	Any
Stringer ori-éntation related to skin orientation	Any	Stringer must be aligned along the skin 0° ply	Any	Any	Stringer must be aligned along the skin 0° ply
Design freedom of skin stacking sequence	Any	Limited, since skin layers are also stringer layers	Any	Any	Limited, since skin layers are also stringer layers
Skin thickness modification potential	Any	Any	Any	Any	Any
Stringer bar thickness modification potential	Impossible	Impossible	Possible	Possible	Impossible
Glass fabric integration	Costly	Non-critical	Difficult under stringer feet	Costly	Non-critical
Edge and stiffness drop at stringer feet	Present	None	Present	Present	None
Rib attachment	Integrated	Non-critical	Limited by stringer feet	Limited by stringer feet	Non-critical
Mandrels (tools) needed per stringer	Umpteen to hundreds	1	3 per stringer family	2 per stringer family	None
Stringer positioning and tolerances	Non-critical	Non-critical	Costly	Non-critical	Non-critical
Number of autoclave cycles	1	1	1	2	1

Questions

1. How would you define „Aeroelastic Tailoring"?
2. What is the root cause of the torsion of a back sweep wing?
3. How will a manoeuvre with increasing lift influence the elastic deformation of a back sweep wing?
4. Name some design possibilities to influence the bending-torsion behaviour of wings for aluminium and CFRP.
5. How will a cantilever made of an anisotropic, orthotropic CFRP laminate deflect, if it is loaded in the centre line of its free end?
6. How will a cantilever made of a non-orthotropic CFRP laminate deflect, if it is loaded in the centre line of its free end?
7. Name some important features of a commercial aircraft which can be influenced by aeroelastic tailoring.
8. What is particularly important when designing CFRP structures with non-orthotropic laminates?
9. How will compression strength of a CFRP laminate (50/40/10) be influenced if the $0°$ fibres are rotated $+15°$?
10. In which direction would you rotate $0°$ layers of a multiaxial CFRP wing skin laminate in order to achieve a higher manoeuvre load alleviation of a back sweep wing?
11. Name a few different CFRP design schemes for wing skins.
12. Why can it be difficult to assemble straight pultruded stringer on wet (non-cured) wing skin? What could be done in order to solve the problem?
13. How would you manufacture U-shaped prepreg stringer profiles? Describe the main steps.
14. Which mechanism is important during L- or U-shaped CFRP profile forming?
15. What has to be considered in terms of stringer orientation if U-shaped CFRP stringer profiles are bonded to a CFRP skin?
16. Name some possible reasons for the "spring-in" effect of CFRP.
17. Name advantages and disadvantages of an integral skin/stringer panel design, in which wet (non-cured) prepreg stringer profiles are co-cured with wet (non-cured) prepreg skin.
18. Name advantages and disadvantages of an integral skin/stringer panel design, in which hard (cured) prepreg stringer profiles are co-bonded to wet (non-cured) prepreg skin.
19. Name advantages and disadvantages of an integral skin/stringer panel design, in which wet (non-cured) prepreg stringer profiles are co-bonded to wet (cured) prepreg skin.
20. Would you expect any kind of inner laminate quality issues if the thermal expansion of aluminium tools is used for the compaction of prepreg stringer profiles during panel manufacturing?

References

1. Breuer, U.: Aeroelastic Tailoring. DGLR Jahrestagung Proceedings (Jahrbuch 1998 Band II), Bremen, Germany, 5–8 October 1998
2. Shirk, M.H., et al.: Aeroelastic tailoring – theory, practice and promise. J. Aircraft **23**(1), 6–18 (1985)
3. Zeitschrift für Luftschiffahrt Nr. 12 (1891)
4. NN: Sherpa Takes the Air, Flight and Aircraft Engineer. **LXIV**(2334), 67–80 (1953)
5. Barnes, C.H.: Shorts aircraft since 1900. Putnam/Aero Publishers, London/Fallbrook, CA (1967)
6. Chan, C.: Flugzeug Prototypen, Vom Senkrechtstarter zum Stealth-Bomber. Motorbuchverlag, Stuttgart (1992)
7. NN: Pterodactyl to Sherpa, Flight and Aircraft Engineer Bd. LXIV, Jahrg. 1953, Heft 20, S. 680–681
8. McCullers, L.A., et al.: Dynamic Characteristics of Advanced Filamentary Composite Structures, vol. II – Aeroelastic Synthesis Procedure Development. AFFDL-TR-73-111, September 1974
9. Waddoups, M.E., et al.: Composite Wing Transonic Improvement, vol. I – Composite Wing Aeroelastic Response Study. AFFDL-TR-71-24, December 1972
10. Waddoups, M.E., et al.: Composite Wing Transonic Improvement, vol. II – Advanced Analysis Evaluation. AFFDL-TR-71-24, December 1972
11. Naberhaus, J.D., Waddoups, M.E.: Dynamic Characteristics of Advanced Filamentary Composite Structures, vol. III – Demonstration Component Program. AFFDL-TR-73-111, September 1974
12. Maske, E.B.: Wing/Inlet Composite Advanced Development. AFFDL-TR-76-88, September 1976
13. Lynch, R.W., et al.: Aeroelastic Tailoring of Advanced Composite Structures for Military Aircraft. AFFDL-TR-76-100, April 1977
14. Brown, L.E., et al.: Aeroelastically Tailored Wing Design. Proceedings "Evolution of Aircraft Wing Design Symposium", pp. 141–146, Dayton, OH, March 1980
15. Forsch, H.: Advanced Design Composite Aircraft (ADCA) Study, vol. I. AFFDL-TR-76-97, November 1976
16. Photo by: NASA, Photo Number: EC87-0182-14; This file is in the public domain because it was solely created by NASA. NASA copyright policy states that "NASA material is not protected by copyright unless noted". (See Template:PD-USGov, NASA copyright policy page or JPL Image Use Policy)
17. Warwick, G.: Forward-Sweep Wing Technology. Flight International, 16 June 1984, pp. 1563–1668
18. NN: X-29 Flies. Flight International, 22 and 29 December 1984, p. 1664
19. NN: X-29 Flight. Flight International, 5 January 1985, p. 10
20. Internet: X-29 Fact Sheet (NASA)
21. Ransom, S.: Zwischen Leipzig und der Mulde, Flugplatz Brandis 1935–1945, p. 73. Stedinger Verlag, Lernwerder (1996). ISBN 3-927697-09-5
22. NN: Im Rückblick HFB320, Internet
23. Triplett, W.E.: Aeroelastic Tailoring Studies in Fighter Aircraft Design. Proceedings 20th Structures, Structural Dynamics and Materials Conference, pp. 72–78, April 1979
24. Triplett, W.E.: Aeroelastic Tailoring of a Forward Swept Wing and Comparison with 3 Equivalent Aft Swept Wings. Proceedings AIAA/ASME/ASCE/AHS 21st Structures, Structural Dynamics and Materials Conference, pp. 754–760, May 1980
25. Borland, C.J., Gimmestad, D.W.: Aeroelastic Tailoring of High Aspect Ratio Composite Wings in the Transonic Regime. AGARD Conference Proceedings 56th Meeting of the SAP, London, UK, 11–12 April 1983

26. Gimmestad, D.: Aeroelastic Tailoring of a Composite Winglet for KC-135. Proceedings 22nd Structures, Structural Dynamics and Materials Conference, pp. 373–376, April 1981
27. Hustedde, C.L., et al.: Further Investigations of the Aeroelastic Tailoring for Smart Structures Concept. Proceedings 38th AIAA/ASME/ASCE/AHS/ASC Structures, Structural Dynamics & Materials Conference, Kissimmee, FL, USA, 7–10 April 1997, pp. 531–537
28. McGowan, A.-M.R., et al.: Results of Wind Tunnel Testing from the Piezoelectric Aeroelastic Response Tailoring Investigation. Proceedings 3ih AIAA Conference Salt Lake City, USA, 15–17 April 1996, AIAA-96-1511-CP, pp. 1722–1732
29. Jackson, P., Bushell, S., Willis, D., Munson, K.: Lindsay Peacock: Jane's all the world's aircraft 2011–2012. Published by Jane's information group, ISBN-10: 0710629559
30. Sensburg, O., et al.: Gust Load Alleviation on Airbus A-300. Proceedings 13th Congress of the International Council of the Aeronautical Sciences, August 1982
31. Schweiger, J., et al.: Aeroelastic Tailoring for Flutter Constraints. MBB Report LKE291-S-PUB-286, München (D), 6 October 1986
32. Schwochow, J.: Statische aeroelastische Untersuchung eines Flugzeugentwurfs und Optimierung der Kohlefaserverbundbauweise des Tragflügels, DGLR Aeroelastik Tagung Göttingen (D), 29–30 June 1998
33. Piening, M.: Ein Strukturkonzept für Tragflügel in Faserverbundbauweise mit Verformungskopplungen. DGLR Aeroelastik Tagung Göttingen (D), 29–30 June 1998
34. Dugas, M.: Ein Beitrag zur Auslegung von Faserverbundtragflügeln im Vorentwurf. Dissertation, Universität Stuttgart (2002)
35. Breuer, U., Müller, J.: Method of fabricating a stringer stiffened shell structure using fiber reinforced composites. United States Patent US 6,306,239

Further Reading

1. Collar, A.R.: The first fifty years of aeroelasticity. Aerospace 5(2), 12–20 (1978)
2. Krone, N.J.: Divergence elimination with advanced composites. Ph.D. thesis, University of Maryland, College Park, MD (1979)
3. Lockenhauer, J.L., Layton, G.P.: RPRV research focus on HIMAT. Astronaut. Aeronaut. 14, 36–41 (1976)
4. Price, M.A.: HIMAT Structural Development Design Methodology. NASA CR-144886, October 1979
5. Chant, C.: Flugzeug-Prototypen, pp. 118–119. Motorbuchverlag, Stuttgart (1994)
6. Wilkinson, K.: FASTOP A flutter and strength optimisation of advanced aircraft structures. J. Aircraft 14, 581–587 (1977)
7. Schneider, G., et al.: Structural Optimisation of Advanced Aircraft Structures. Proceedings 12th Congress of the International Council of the Aeronautical Sciences, October 1980
8. Rogers, W.A., et al.: Validation of Aeroelastic Tailoring by Static and Aeroelastic Flutter Tests. AFWAL-TR-81-3160, September 1982
9. Rogers, W.A., et al.: Design, Analysis and Model Tests of an Aeroelastically Tailored Lifting Surface. AIAA-81-1673, August 1981
10. Wilkinson, K., Rauch, F.: Predicted and Measured Divergence Speeds of Advanced Composite Forward Swept Wing Model. AFWAL-TR-80-3059, July 1980
11. Ellis, J.W., et al.: Structural Design and Wind Tunnel Testing of a Forward Swept Fighter Wing. AFWAL-TR-80-3073, July 1980
12. Schweiger, J., et al.: Aeroelastic Problems and Structural Design of a Tailless CFC-Sailplane. Proceedings International Symposium on Aeroelasticity and Structural Dynamics, Aachen (D), April 1985
13. Cook, E.L., et al.: Weight Comparison of Divergence Free Tailored Metal and Composite Forward Swept Wings for an Executive Aircraft. Proceedings International Conference, University of Bristol, UK, 24–26 March 1982

14. Booker, D.: Aeroelastic Tailoring for Control and Performace – Are Requirements Compatible? AGARD Conference Proceedings No. 319 FMP Symposium Florence, Italy, 5–8 October 1981
15. Bohlmann, J.D.: Application of Analytical and Design Tools for Fighter Wing Aeroelastic Tailoring. AGARD Report 784, AGARD Structures and Materials Panel, Bath, UK, 29.4.–3.5.1991
16. Weisshaar, T.A.: Tailoring Methodology for Aeroelastic Stability and Lateral Control Enhancement. DGLR Report 85-02, 2nd International Symposium on Aeroelasticity and Structural Dynamics, Aachen, Germany, 1–3 April 1985
17. Patil, J.M.: Aeroelastic Tailoring of Composite Box Beams, AIAA 97-0016. Proceedings 35th Aerospace Sciences Meeting & Exhibit, Reno, NV, USA, 6–10 January 1997, pp. 1–9
18. Librescu, L., Chandiramani, N.K.: Aeroalastic Tailoring of Laminated Composite Shear Deformable Fiat Panels Exposed to Supersonic Coplanar Gas Flow of Arbitrary Direction. DGLR Report 91-110, pp. 404–410
19. Chen, R., Butler, R.J.: Divergence of Forward Swept Wings. Proceedings International Conference Forward Swept Wing Aircraft, University of Bristol, UK, 24–26 March 1982, pp. 11.61–11.69
20. Chen, G.-S., Dugundji, J.: Experimental aeroelastic behavior of forward-swept graphite/epoxy wings with rigid-body freedom. J. Aircraft 24(7), 454–462 (1987). doi:10.2514/3.45501
21. Weisshaar, T.A.: Aeroelastic Tailoring of Forward Swept Composite Wings. AIAA-80-0795
22. Weisshaar, T.A.: Creative Uses of Unusual Materials. AIAA 87-0976-CP. Proceedings AIAA/ASME/ASCE/AHS 281h Structures, Structural Dynamics and Materials Conference, Monterey, CA, 6–8 April 1987
23. Kerr, R.I., Thompson, D.: Automated Structural Optimisation at Warton. ICAS-86-3-1-3
24. Dillenius, M.F.E., et al.: Aeroelastic Tailoring Procedure for Controlling Hinge Moments, AIAA-88-0528. AIAA 26th Aerospace Sciences Meeting, Reno, Nevada, USA, 11–14 January 1988
25. Dillenius, M.F.E., et al.: Aeroelastic Tailoring Procedure to Optimise Missile Fin Centre of Pressure Location, AIAA-92-0080. 30th Aerospace Sciences Meeting & Exhibit, Reno, Nevada, USA, 6–9 January 1992
26. Schneider, G., et al.: Aeroelastic tailoring Validation by Wind Tunnel Testing. MBB-Report FE291-S-PUB-363, München (D), April 1989
27. Bohlmann, J.D., Scott, R.C.: A Taguchi Study of the Aeroelastic Tailoring Design Process. AIAA-91-1041-CP
28. Love, M.H., Bohlmann, J.D.: Aeroelastic Tailoring in Vehicle Design Synthesis. AIAA-91-1099-CP
29. Ehlers, S.M., Weisshaar, T.A.: Static Behaviour of an Adaptive Laminated Piezoelectric Composite Wing. AIAA-90-1078-CP
30. Livine, E., et al.: Integrated structure/control/aerodynamic synthesis of actively controlled composite wings. J. Aircraft 30(3), 387–394 (1993)
31. Garfinkle, M.: Twisting smartly in the wind. Aero. America 32(7), 18–20 (1994)
32. Piening, M.: Die statische Aeroelastizität des anisotropen Tragflügels. DFVLR Report IB 131-84157
33. Piening, M.: Der aeroelastische Flügel: Leistungsgewinn durch Einsatz anisotroper Werkstoffe. DFVLR Nachrichten Heft 42 (Juni 1984)
34. Layton, J.B.: Aeroelastic and Aeroservoelastic Tailoring with Geometry Variables for Minimising Gust Response of Cantilevered Finite Span Wing, pp. 1086–1103. AIAA-96-1442-CP
35. Giese, C.L., et al.: An Investigation of the Aeroelastic Tailoring for Smart Structures Concept, pp. 2274–2278. AIAA-96-1575-CP
36. Sanderson, R.: Immer größer, schwerer, schneller, weiter. Die Aerodynamik vor neuen Herausforderungen. EF-Kolloquium 1997 DASA-Airbus, Bremen

Chapter 9
New Developments

Abstract CFRP was the prime choice for the airframe of the latest aircraft of Boeing and Airbus. However, modern aluminium alloys were developed with improved cost-performance relationships, and, in addition, todays CFRP manufacturing technology is not fully ready for high production rates of short- and medium-range aircraft. Furthermore, the light weight design potential of today's CFRP airframe is heavily penalised by additional system installation effort, which is necessary due to the poor electrical conductivity of CFRP compared to aluminium or fibre metal laminate such as Glare®. For the next generation of aircraft, it is necessary to reduce material cost *and* airframe manufacturing cost, and to improve the added product value by additional functions such as electrical conductivity. This chapter reveals latest R&D results to enable cost-effective thermoplastic composite airframe structures with blends, more electrically conductive and more damage tolerant composites with metal fibre reinforcements and up-stream research for different use cases of morphing structures by means of shaped memory alloy wire integration.

Keywords Material cost • Polymer blends • PEEK • PPS • PES • Glass transition temperature • Storage modulus • Tensile strength • Young's modulus • Elongation at break • Crystallinity • Annealing • Electrical conductivity • Specific electrical resistance • Metal fibre reinforcement • Damage tolerance • Failure strain • Energy absorption • Multi-functional material • Shape memory alloy • One-way effect • Two-way effect • Adaptive winglet • Adaptive vortex generator • Adaptive chevron • Adaptive trailing edge • Slat-cove filler • Active airfoil

Thermoplastic Blends: Low Cost But High Performance

Tim Krooß and Ulf Breuer

Fibre reinforced thermoplastics provide several advantages towards comparable composites with thermosetting polymer matrices. For industrial processing the shorter cycle time is one of the most interesting arguments to consider thermoplastics. Ease of material storage, advanced joining technologies such as thermoplastic-welding and recycling underline additional advantages [1]. However, for primary airframe structures, thermoset resin systems are still dominating, and the share of high performance thermoplastic composites within airframe structures is still relatively small (see Chap. 4).

Thermoplastic applications include polycarbonates (PC), polyamides (PA), polyphenylene sulfides (PPS), polyether imides (PI) and polyether ether ketones (PEEK) [2], but only PEEK and PPS are preferred for primary structure applications. PEEK/carbon fibre materials demonstrate remarkable weight savings compared to aluminium alloys [3, 4]. Additionally, excellent chemical resistance and good high temperature performance make PEEK very attractive for airframe structures.

PEEK, however, is not only the best performing thermoplastic polymer, but also the most expensive one. The price of virgin PEEK can exceed 70 €/kg, and the price of carbon fibre reinforced PEEK is between approx. 100 and 200 €/kg, depending on the type of fibres. This has a negative impact on the total manufacturing cost the composite parts.

The reduction of material cost must be seen as a key enabler for more thermoplastic composite applications in the airframe.

Polymer Blends

The possibility of mixing different polymers, leading to improved physical, mechanical and thermo-mechanical properties, offers a broad field of applications for both commodity (cheap) and high performance polymers.

Polymer blends are usually divided into two groups: miscible and immiscible blends [5, 6]. While miscible polymer blends show a homogenous and stable behaviour and their properties tend to be intermediate between the properties of the constituents, the properties of immiscible blends are strongly affected by their phase morphologies [7]. The main influence parameters for morphology formation and therefore for the properties are their relative viscosity, interfacial tension and processing conditions. In some cases, immiscible blends have an unstable morphology and tend to segregate over time, which outlines the necessity of compatibilisation to make them suitable materials for commercial applications [8].

The different types of polymer blends are related to various fields of application and the desired blend characteristics, respectively. A common process for preparing blends is mechanical melt mixing. This generally solvent-free process is easily scalable to various sizes. The polymers are heated above their melting point and then mixed under the influence of shear stress. Commercially, thermoplastic polymer blends are produced via extrusion processing. This is one of the most important processes for polymer blend production [9].

Depending on the specific polymer combination and morphological needs, single or multi screw extruders can be used. Multiple screws allow for specific regulation of shear stress by modification of screw elements. The shear stress as well as temperature, viscosity and weight ratio of the inserted polymers are main parameters to influence the morphology of the blend and thus its overall performance.

Blend Morphologies

The formation of certain morphology in immiscible blends can be controlled by the manufacturing process. Several types of morphologies are possible, resulting in different blend properties. Disperse morphology can, for example, be used for surface modification. Laminar structures increase the barrier properties and co-continuous morphologies can elevate the impact resistance (toughness) of the resulting blend and its composites. The formation of fibrils generates anisotropic properties and strengthens the matrix material due to high molecular orientation and increased crystallinity [7, 10].

The theoretical model for these morphology formations is based on the collision-induced breakup and coalescence mechanisms in the polymer melt. The driving factor for these phenomena is the so-called capillary number, the proportion of deformation and interfacial tension of a polymer melt phase or droplet. If the capillary number falls below a critical value, breakup of phase droplets can occur and dispersion becomes possible [11]. Figure 9.1 illustrates the correlation of phase morphology and blend composition. With growing concentration of dispersed droplets, the coalescence increases. Under shear, the elongation of grown droplets is enhanced and the morphology changes in the region of phase inversion towards a co-continuous morphology.

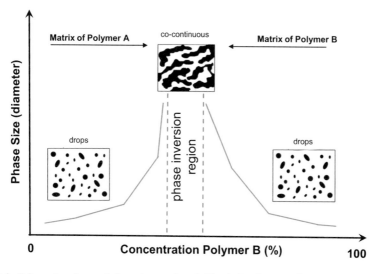

Fig. 9.1 Schematic of morphology types of melt-blended polymer-polymer compounds as a function of blend composition according to [7, 12]

Within the production process the viscosity of the selected polymers can directly be influenced by temperature and shearing conditions, and thus an adjusted formation of morphology is possible. This technology offers the possibility of tailoring thermoplastic matrix materials related to their targeted application. Polymer blends for continuously fibre reinforced composites and high performance applications are still rarely investigated.

PPS-PES Blend Developments

In recent investigations, thermoplastic polymer blends have been developed and analysed as potential future candidates for the high-performance thermoplastics PEEK and PPS [13]. For this purpose, the combination of PPS with PES was investigated. In former studies those two polymers were found to be poorly or only partially miscible [14, 15]. This means that the mixing will not yield in blend properties to be expected of a single phase material [9]. A microscopic morphology of droplets or spheres will be formed, depending on the mass ratio and the processing conditions.

The target was to generate a dispersion of PES in a PPS matrix. PPS has a superior chemical resistance and a high Young's modulus. PES on the other hand is well known for its thermal resistance due to its high glass transition at approx. 220 °C and its high ductility at high strength, which qualifies this polymer as a potential blend partner for PPS. The main specifications to be met by the new blends were:

- sufficient chemical resistance towards liquid media (provided by the PPS matrix or a co-continuous morphology)
- thermal stability comparable to PEEK within the typical operating temperature range
- maximum mechanical strength and modulus in relation to the neat polymers

The conclusions of this work shall provide the basic knowledge for blend processing via film and melt-spinning extrusion for further investigations and potential future applications in CFRP.

In the first step, the definition of adequate blend compositions by means of rheological analysis of different polymer grades was necessary. In dependence of the viscosity ratios of the selected PPS and PES grades, a rheological profile of several polymer combinations was defined, providing advice for a successful morphology formation. It was targeted to obtain minimum viscosity ratios of the polymer grades, being reported to easier enable droplet breakup and thus the formation of fine dispersed droplet morphology [16, 17]. Additional factors influencing the microstructure were the processing shear rate, processing temperature and weight fractions of the polymers. As a result of the detailed rheological analysis, the selection of one polymer grade for each type was possible. The

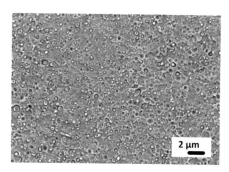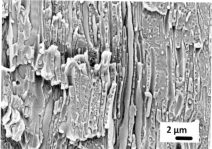

Fig. 9.2 *Left*: Cryo-fractured cross-section of 30 wt% PES reinforced PPS (dispersion), *right*: cryo-fractured cross-section of 50 wt% PES reinforced PPS (mixed/co-continuous)

processing conditions for injection moulding and the film extrusion process were derived from the temperature and shearing optima of the viscosity measurements.

Scanning electron microscopy (SEM) of fracture surfaces of the injection moulded samples showed that a generally fine dispersion of 15 wt% and 30 wt% PES in the PPS was created (Fig. 9.2, left), leading to a quasi-homogeneous morphology, while the inversion from a PPS matrix towards a PES or co-continuous matrix occurred at 40 wt% (Fig. 9.2, right). By addition of a compatibiliser, this phenomenon was significantly inhibited. The composition with 47 wt% PES showed the sophisticated inversion of the PPS matrix, which could also be suppressed by the presence of a compatibiliser.

In conclusion, by means of this SEM analysis, characteristic compositions could be identified in which the blends' morphologies experience significantly changes. Hence mechanical, thermo-mechanical and chemical properties of the moulded samples should vary as a function of their specific morphology and polymer fractions.

A first performance assessment of these high-performance polymer blends was the analysis of their thermo-mechanical behaviour. Neat injection moulded samples as well as samples aged in methyl ethyl ketone (MEK) and annealed samples were tested in dynamic mechanical thermo analysis (DMTA) experiments and compared to non-blended virgin PPS and PEEK samples, Fig. 9.3.

The results demonstrate the high potential of adequately tailored polymer blends and their ability to substitute expensive high performance thermoplastics or even exceed their properties.

At low temperatures, the high modulus PPS fraction is dominating the thermo-mechanical behaviour of the blends until its glass transition is reached at approx. 90 °C. With the PES fractions shown here, the modulus decreased in this area. However, a well formed plateau at higher temperatures with significantly better performance than PEEK and PPS was observed. Depending on the composition, this plateau varies in height. An additional annealing step of blended PPS samples with 15 wt% PES at 250 °C for 7 days increased the thermo-mechanical

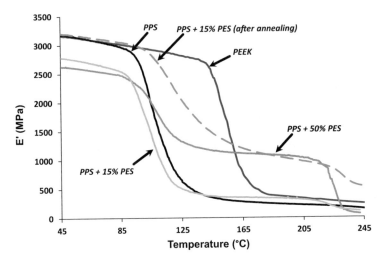

Fig. 9.3 Storage modulus of different PPS/PES blend compositions as function of temperature compared to neat PPS and PEEK

performance over the whole temperature range, since the degree of crystallinity was enhanced from approx. 13 % to approx. 80 %.

Regarding the chemical resistance, it could be demonstrated by DMTA measurements that compositions up to 30 wt% PES exhibit good performance after being aged in MEK for approx. 200 h. Higher PES fractions led to a significant drop of properties due to loss of the resistant PPS phase, protecting the samples inner PES structure.

Mechanical testing at room temperature of PPS-PES blends demonstrated good performance. Table 9.1 shows the tensile properties of samples of two different blend compositions in comparison to the neat virgin thermoplastics PEEK, PPS and PES. The dog bone samples according to DIN EN ISO 527-2 were produced by film stacking and consolidation via hot pressing.

While the tensile strength of the blends was not significantly affected by the weight fraction of the components, Young's modulus shows a remarkable dependence on the PES amount. At low PES concentration, the blend was more influenced by the high modulus PPS component, while higher PES fractions lead to slightly increased ductility.

In this specific case of a semi-crystalline polymer (PPS) and an amorphous polymer (PES) mixture, another important role could be attributed to the crystallisation kinetics of the blend. As illustrated in Fig. 9.4, the degree of crystallinity was significantly influenced by the thermal treatment as well as by the composition of the blends. Since the PES fraction is completely amorphous, the crystallisation kinetics of the PPS fraction was noticeably inhibited, which has also been investigated in former studies for these types of blends [9].

Especially in the case of PPS, the degree of crystallinity has significant influence on its impact strength and fracture toughness. In [19] and [20] it is stated that impact

Table 9.1 Comparison of mechanical properties (mean values) of PPS-PES blends with high performance thermoplastics

Property	PEEK	PPS	PES	PPS/PES 85/15	PPS/PES 50/50
Tensile strength (MPa)	100	90	90	80	81
Young's modulus (MPa)	3700	4200	2700	3650	3250
Elongation at break (%)	40	8	>25	2.6	3.4

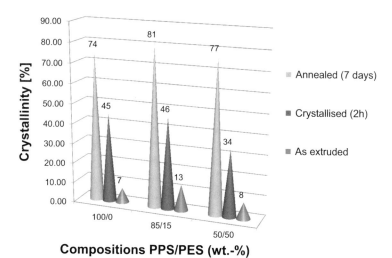

Fig. 9.4 Crystallinity of PPS-PES blend films after extrusion (virgin PPS film), after crystallisation at 204 °C for 2 h and after annealing at 250 °C for 7 days. Calculated from DSC measurements, 100 % values taken from [18]

strength as well as fracture toughness decrease with increasing crystallinity. Important composite properties which are influenced by matrix properties, such as the impact damage tolerance, must be taken into account for the definition of the blend composition and the thermal treatment of the blend material. In the case of PPS/PES blends, the crystallinity will also affect other physical and chemical matrix properties and must be further investigated in future studies.

Conclusion and Outlook

The principle of blending technology offers promising perspectives for high-performance applications with PPS-PES. Tailoring of blend properties through controlled process adjustment is a cost-effective possibility to obtain superior material properties. Further research will concentrate on resulting blend properties in endless carbon fibre reinforced material, with a special focus on phase separation and crystallisation behaviour.

Metal and Carbon United: Electrical Function Integration

Benedikt Hannemann, Ulf Breuer, Sebastian Schmeer, Sebastian Backe, and Frank Balle

Pushed by the need for further mass reduction and structural performance improvements, polymer composites have been increasingly used for airframe applications. CFRP was the primary choice for wing, tail plane and fuselage for Boeing's B787 or Airbus' A350 [21, 22].

To contribute to further mass reduction of next generation airframes, efforts must focus on more affordable airframe structures and material multi-functionality [23]. Since the electrical conductivity of CFRP is poor compared to conventional aluminium or Glare®, additional metal masses are needed in today's CFRP aircraft to fulfil the necessary electrical functions. Former research attempts concentrated on modifying the polymer matrix systems [53]. By introduction of conductive particles such as carbon nano tubes, the specific conductivity of CFRP could be enhanced [24]. However, a sufficient level of conductivity, which would guarantee electrical function integration for the modified CFRP similar to that of aluminium alloys and GLARE® airframe structures, could not be demonstrated.

The impact damage tolerance of thin-walled CFRP structures has gradually been improved by the addition of polymer toughening agents. Thermoplastic polymers and rubber particles were introduced in epoxy resin systems in different ways for prepregs, enabling substantial improvements of fracture toughness and residual strength. However, even for CFRP airframe structures made with the latest generation of toughened prepreg systems, damage tolerance against probable impact events can be the limiting design factor.

Metal Carbon Fibre Composites

Unlike former research attempts dealing with modified polymer matrix systems, the improvement of important composite properties as well as the integration of electrical functions by means of reinforcing metal fibre incorporation is a promising new approach. Research results in [25–28] show significant improvements in terms of energy absorption, fail safe behaviour and structural integrity, when classical composites were reinforced by steel fibres.

The metal fibres can be distributed homogenously in the composite (homogenised layer approach) or locally concentrated in separate layers (separated

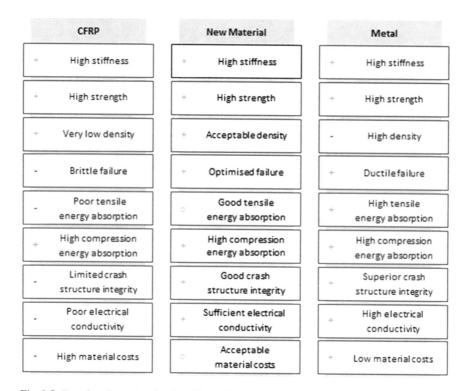

CFRP		New Material		Metal	
+	High stiffness	+	High stiffness	+	High stiffness
+	High strength	+	High strength	+	High strength
+	Very low density	+	Acceptable density	-	High density
-	Brittle failure	+	Optimised failure	+	Ductile failure
-	Poor tensile energy absorption	o	Good tensile energy absorption	+	High tensile energy absorption
+	High compression energy absorption	+	High compression energy absorption	+	High compression energy absorption
-	Limited crash structure integrity	+	Good crash structure integrity	+	Superior crash structure integrity
-	Poor electrical conductivity	+	Sufficient electrical conductivity	+	High electrical conductivity
-	High material costs	o	Acceptable material costs	+	Low material costs

Fig. 9.5 Benefits of metal and carbon fibre reinforced composite

layer approach). The basic idea of this hybrid material concept is to use the metal fibre for electrical and structural load-bearing functions, Fig. 9.5.

The increased density of the hybrid composite must be overcompensated by the benefit of eliminating additional electrical system installation items and a reduced minimum wall thickness, exploiting its improved electrical conductivity and damage tolerance.

A steel fibre volume content of 20 % can be considered as a reasonable upper limit. At this volume content, the composite density reaches the density of aluminium (approx. 2.78 g/cm^3, depending on the specific aluminium alloy), but, more importantly, the metal content should be limited by the amount of additional metal needed today for CFRP airframe structures in order to fulfil all necessary electrical functions. The use of thin metal filaments is advantageous for design reasons. Different to fibre-metal-laminates such as GLARE® or ARALL®, the fibre based approach enables stress tailored composite design and complex shaped structures. Furthermore, fully automated manufacturing technologies can be explored by means of processes already available for CFRP (e.g. weaving processes for non-crimped fabric manufacturing, automated fibre placement, tape laying, RTM etc.). However, potential metal fibres have to meet various requirements, in particular superior electrical conductivity, distinctive failure strain, high strength, low

	Aluminum	Copper	Steel
Density	+	-	-
Ultimate tensile strength	o	-	+
Failure strain	+	+	+
Thermal expansion	-	+	o
Electr. conductivity	+	+	o
Therm. conductivity	o	+	-
Corrosion resistance	-	+	+
Costs	o	-	+

Fig. 9.6 Comparison of potential metal fibre materials

density, corrosion resistance, appropriate thermal expansion, availability and low costs, Fig. 9.6.

Aluminium fibres are distinguished by superior weight specific electrical and mechanical properties. However, in direct contact with carbon fibres and electrolytes, aluminium tends to ineligible galvanic corrosion. This is of no relevance for stainless steel fibres, since they are commercially available with a wide range of mechanical properties and appearance. The properties depend not only on the alloy composition and the heat treatment, but also on the filament diameter. Typically, decreasing the diameter leads to a lower ultimate strain to failure. Compared to a standard high tenacity ex-PAN carbon fibre, the electrical conductivity of soft annealed stainless steel fibres (AISI 304) is about 24 times higher. At the same time, the deformation capability can be 16 times higher, while the ultimate tensile strength (UTS) is five times lower. The stiffness for tensile load of carbon and stainless steel fibres is almost comparable [29].

Due to less alloying elements, unalloyed steel fibres have an even higher specific electrical conductivity, but worse mechanical properties. By using nickel or copper cladding, the electrical conductivity of steel fibres can be further enhanced up to a factor of 140 and 390, respectively.

	Carbon fibre	Stainless steel fibre
Density / g/cm^3	1.77	7.95 ± 0.01
Tensile modulus / GPa	240 ± 3	176 ± 7
Ultimate tensile strength / MPa	4,806 ± 125	897 ± 2
Failure strain / %	1.85 ± 0,04	32.31 ± 2.01
Electr. resistivity / Ω·m	1.6×10^{-5}	$(6.97 \pm 0.02) \times 10^{-7}$
Filament diameter	5	60.0 ± 0.4
Roving size	12k	7

Fig. 9.7 Properties of applied carbon fibres (manufacturer data) and stainless steel fibres (own measurements)

Material Preparation

In order to validate the electrical and mechanical properties of carbon and steel fibre reinforced epoxy resin, unidirectional laminates were manufactured in a combined process of tape deposition and filament winding technology. Unidirectional layers of pre-impregnated high tenacity carbon fibres were stacked on a flat steel core and wrapped in dry stainless steel fibre bundles. Each bundle consists of seven twisted filaments with a diameter of 60 μm. Both electrical and mechanical properties of the bundles are known from previous tests, Fig. 9.7. Under tensile load, the individual filaments of the twisted steel fibre bundles are inconsistently loaded, causing a relatively low Young's modulus.

The resin required for steel fibre impregnation originates from the bleed of the carbon fibre prepreg layers and resin films, respectively. Additionally, pure carbon fibre reinforced plastic (CFRP) and steel fibre reinforced plastic (SFRP) were prepared as reference material. All laminates were cured at 180 °C for 3 h using autoclave technology and subsequently released from the winding core. By this procedure, six multi-layered laminates with different fibre volume fractions and a cured thickness of approximately 1 mm were manufactured, Fig. 9.8. Finally, specimens were retrieved from the cured plates by means of water jet cutter technology and prepared for subsequent testing.

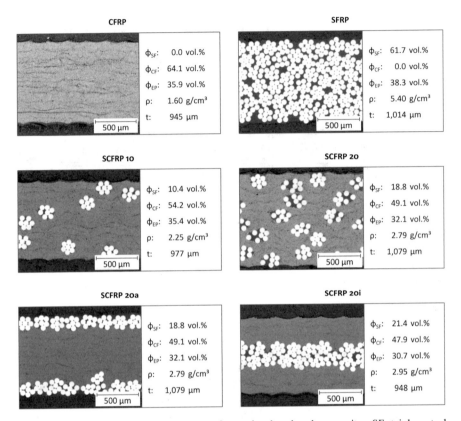

Fig. 9.8 Architecture and properties of manufactured and analysed composites. *SF* stainless steel fibre, *CF* carbon fibre, *EP* epoxy resin, *ρ* density, *t* laminate thickness

Analytical Estimations

The specific electrical conductivity defines the material's ability to conduct electric current. For a conductor with a uniform cross section, the specific electrical conductivity κ is defined as

$$\kappa = S \cdot \frac{l}{A} \qquad (9.1)$$

where

    S is the electrical conductivity

    A is the cross-sectional area

    l is the length of the specimen

The inverse of the specific electrical conductivity κ is called the specific electrical resistance ρ. Based on the electrical properties of its components, the electrical conductivity of a perfect unidirectional endless fibre reinforced polymer can be estimated by means of the rule of mixture. Longitudinal to the fibre direction, the composite can be considered as a parallel circuit of several conductors. The overall conductivity is given by

$$S = \sum_i S_i \qquad (9.2)$$

Regarding the volume fraction of its components φ_i, the mean specific conductivity along the fibre orientation κ can be calculated considering the volume fractions φ_i and the corresponding electrical conductivities κ_i

$$\kappa = \sum_i \kappa_i \cdot \varphi_i \qquad (9.3)$$

This relation for a three-component composite with a constant resin fraction of 40 vol% is shown in Fig. 9.9, assuming an electrical conductivity of 1.43×10^6 S/m for stainless steel fibres and 6.25×10^4 S/m for high tenacity carbon fibres [29].

Following this theoretical approach, a steel/carbon fibre hybrid composite should demonstrate an electrical conductivity of more than five times the conductivity of pure CFRP for a steel fibre fraction of 10 vol% and eight times for a steel fibre fraction of 20 vol%. In the meantime, the density rises from 1.59 to 2.20 g/cm^3 and 2.82 g/cm^3 (for comparison aluminium: 2.71 g/cm^3). Regarding an electrical conductivity of 2.36×10^7 S/m for commercially available copper cladded low carbon steel fibres, the overall conductivity of the hybrid composite can be increased by a factor of 64 and 127, respectively.

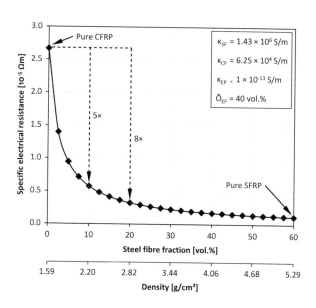

Fig. 9.9 Specific electrical resistance and density of a hybrid composite

Material Characterisation: Mechanical Properties

Monotonic tensile tests longitudinal to the fibre orientation were performed with a hydraulic testing machine of type Zwick Roell HTM 50/20 in accordance to DIN EN ISO 527-5:2010-01. Rectangular specimens were sized 250 mm × 15 mm and provided with 3 mm thick chamfered aluminium end tabs for load introduction. The specimens were clamped with a gauge length of 150 mm and loaded with a crosshead speed of 3 mm/s. The deformation was recorded by a high-speed camera and evaluated by an optical 2D measuring system. Representative stress-strain curves are shown in Fig. 9.10.

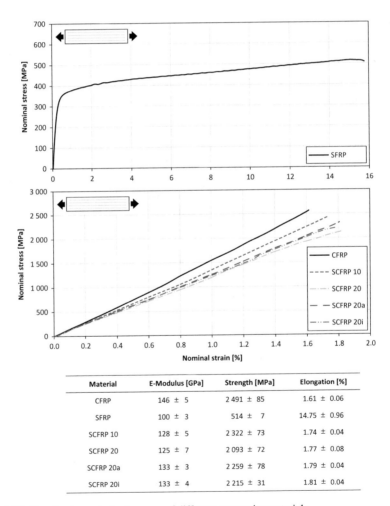

Material	E-Modulus [GPa]	Strength [MPa]	Elongation [%]
CFRP	146 ± 5	2 491 ± 85	1.61 ± 0.06
SFRP	100 ± 3	514 ± 7	14.75 ± 0.96
SCFRP 10	128 ± 5	2 322 ± 73	1.74 ± 0.04
SCFRP 20	125 ± 7	2 093 ± 72	1.77 ± 0.08
SCFRP 20a	133 ± 3	2 259 ± 78	1.79 ± 0.04
SCFRP 20i	133 ± 4	2 215 ± 31	1.81 ± 0.04

Fig. 9.10   Tensile stress-strain diagrams of different composite materials

CFRP and the hybrid composites (SCFRP 10, SCFRP 20, SCFRP 20a, SCFRP 20i) show a linear elastic behaviour with similar strains at failure between 1.61 % and 1.81 %. The mean ultimate tensile strength decreases from 2491 MPa to 2093 MPa, 2322 MPa, 2259 MPa and 2215 MPa. Moreover, the stiffness declines from 146 GPa for pure CFRP to approximately 130 GPa for the hybrid composites. By contrast, pure SFRP shows a pronounced ductile material performance with a yield strength of 349 MPa, a ultimate strength of 514 MPa and an ultimate strain at failure of 14.75 %.

In addition, tensile tests lateral to the fibre orientation were conducted with a conventional testing machine of type Zwick 1474 in dependence on DIN EN ISO 527-5:2010-01. The test samples were sized 150 mm × 15 mm and provided with 1 mm thick chamfered GFRP end tabs. The specimens were clamped with a gauge length of 50 mm and loaded with a test speed of 1 mm/min. Again, the deformation was analysed by an optical 2D measuring system. The results are summarised in Fig. 9.11.

Both CFRP and the hybrid composites show a linear elastic behaviour. CFRP demonstrates an average tensile strength of 80 MPa and an ultimate elongation of 0.92 %. The incorporation of steel fibres to the composites increases the stiffness but lowers the strain at failure, which also causes a decrease of the tensile strength. This aspect is more pronounced for the hybrid composites with the concentrated steel fibres (SCFRP 20a and SCFRP 20i) than for the hybrid composites with the homogenous steel fibre distribution (SCFRP 10 and SCFRP 20).

Flexure tension tests were carried out on a conventional testing machine (Zwick 1474). For this purpose, flat rectangular coupons (125 mm long and 15 mm wide) were clamped at their short edges with a remaining span of 50 mm and centrally loaded by a rounded indenter with a cross head speed of 2 mm/min, Fig. 9.12. The supports were chamfered with 5 mm radius. Fibres were orientated in parallel to the long edge of the specimen. Typical force-displacement curves are given in Fig. 9.13. The integrals of the curves yield the absorbed deformation energy.

Pure CFRP shows a brittle material performance. Failure occurs abruptly and singularly with a bearable deformation of 3.58 mm. By contrast, the incorporation of stainless steel fibres into CFRP causes noticeable post failure behaviour. In case of the hybrid composites with homogenously distributed stainless steel fibres, first failure occurs at 3.36 mm (SCFRP 10) and 3.22 mm (SCFRP 20), respectively. Afterwards, the coupons can be further loaded at a reduced level of force. Total failure is reached at an increased deflection of 6.11 mm and 6.33 mm, respectively. An even better material performance can be realised by localising the metal fibres in the core (SCFRP 20i) or top layers (SCFRP 20a) of the hybrid laminate.

Material Characterisation: Electrical Laminate Properties

Electrical conductivity measurements for the laminate in-plane properties were examined by the four point probes method on a LabVIEW-based data logging

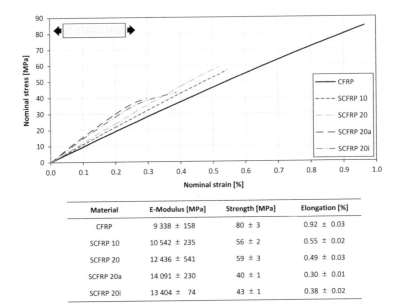

Material	E-Modulus [MPa]	Strength [MPa]	Elongation [%]
CFRP	9 338 ± 158	80 ± 3	0.92 ± 0.03
SCFRP 10	10 542 ± 235	56 ± 2	0.55 ± 0.02
SCFRP 20	12 436 ± 541	59 ± 3	0.49 ± 0.03
SCFRP 20a	14 091 ± 230	40 ± 1	0.30 ± 0.01
SCFRP 20i	13 404 ± 74	43 ± 1	0.38 ± 0.02

Fig. 9.11 Tensile stress-strain diagrams of different composite materials

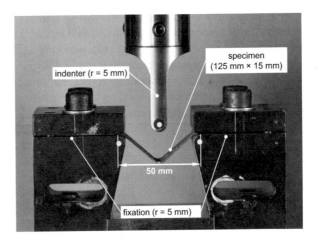

Fig. 9.12 Experimental setup of flexure tension tests

system. For this purpose, direct current (DC) was applied to the short surfaces of 15 mm wide rectangular specimens via copper electrodes. To guarantee a reproducible electrical connection between the measuring device and the test item, the copper electrodes were pressed against the short surface of the specimen with a defined pressure. In addition, the short surfaces of the specimen were treated with a picosecond laser to expose the ends of the steel fibres by removing the surrounding

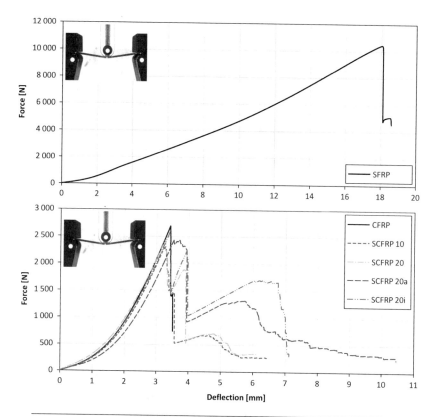

Fig. 9.13 Force-displacement diagrams for combined tension-bending load

Material	Max. Force [N]	Initial failure [mm]	Max. deflection [mm]	Absorbed energy [J]
CFRP	2 715 ± 105	3.41 ± 0.04	3.58 ± 0.12	3.38 ± 0.09
SFRP	9 797 ± 742	17.14 ± 0.89	17.48 ± 1.20	75.96 ± 11.30
SCFRP 10	2 396 ± 93	3.36 ± 0.07	6.11 ± 0.28	4.24 ± 0.21
SCFRP 20	2 406 ± 57	3.22 ± 0.07	6.33 ± 0.36	5.16 ± 0.34
SCFRP 20a	2 467 ± 38	3.79 ± 0.21	10.79 ± 0.47	9.24 ± 1.41
SCFRP 20i	2 568 ± 130	3.45 ± 0.05	7.07 ± 0.26	8.19 ± 0.22

material, Fig. 9.14. Finally, the short faces of the specimen were carefully ground, cleaned by ethanol and brushed with a silver conductive paste.

The surface temperature was monitored during the conductivity experiments by IR-thermography to exclude thermal-induced changes of the electrical conductivity. The experimental setup is shown in Fig. 9.15.

For a period of 10 s a various direct current was applied to the specimen. The reading was repeated for specimen length of 30–90 mm. The voltage was metered for DC stages of 50, 100, 150, 200 and 250 mA. Based on the specific geometry of

Fig. 9.14 SEM image of a
laser treated hybrid
composite's contact face

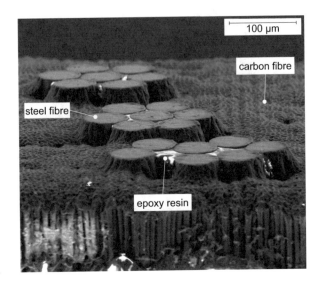

Fig. 9.15 Experimental
setup for electrical
properties:
(1) IR-thermography,
(2) specimen fixture with
copper electrodes,
(3) DC-supply, (4) voltage
metering

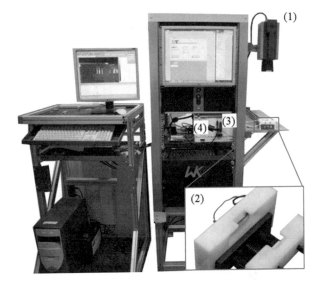

each and every specimen the specific electrical resistance was calculated by Ohm's
law. The results of the conductivity measurements for the different composite
configurations are summarised in Fig. 9.16.

Fig. 9.16 Specific electrical resistance of unidirectional CFRP, SFRP and SCFRP

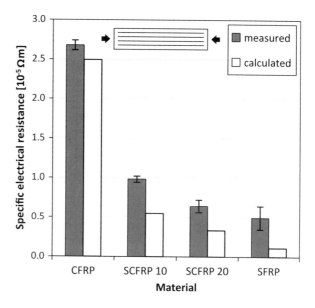

Material	Specific resistance [10^{-5} Ωm]	
	calculated	measured
CFRP	2.50	2.68 ± 0.06
SCFRP 10	0.55	0.90 ± 0.04
SCFRP 20	0.33	0.64 ± 0.08
SFRP	0.11	0.49 ± 0.15

Results and Discussion

In case of pure tensile load in parallel to the fibre orientation, both CFRP and the hybrid composites show brittle material performances with comparable ultimate strains to failure. Despite the incorporation of highly ductile stainless steel fibres, pronounced post failure behaviour could not yet be demonstrated. Due to the homogenous stress state and material deformation, the carbon fibres store elastic energy on the entire length of the specimen. At the time of their failure, this energy is abruptly released and transferred by the matrix to the ductile steel fibres. However, this energy transfer is limited very locally at the point of failure. By this means, the potential ductility of the steel fibres can only be addressed on a very small scale, causing merely a slight macroscopic elongation. The total elongation at break coincides with the strain at failure of the carbon fibres. Improvements are expected by higher steel fibre fractions (i.e. less elastic energy that must be absorbed by a bigger amount of plastifying steel fibres), but would cause higher material densities. This assumption is supported by the test results of pure SFRP, which demonstrate a pronounced plastic deformation. Furthermore, a reduced fibre-resin-adhesion could enable unhindered deformation of the metal fibres over more extensive distances, but would certainly impact other important properties.

In case of tensile load lateral to the fibre orientation, the isotropic behaviour of the steel fibres significantly affects the mechanical properties of the hybrid composites. Compared to the (anisotropic) carbon fibres, the stainless steel fibres offer a much higher stiffness, which globally causes a higher lateral stiffness of the hybrid composites. However, the high stiffness of the steel fibres also provokes an increase of the resin deformation and therefore lower elongation at failure of the composite. Both phenomena enhance with higher steel fibre fractions. The effect of the strain increase promotes crack initiation at low composite strains in the region of pure SFRP. In case of the unidirectional reinforced composites SCFRP 20a and SCFRP 20i, unhindered crack growth causes sudden failure at low composite deformations.

The potential of incorporated steel fibres for an improved fracture behaviour of the hybrid composite becomes apparent in the flexure tension tests, Fig. 9.17. As in many applications, load is very locally introduced. High stress peaks occur in the loaded regions, while stress states in the rest of the specimen are significantly lower. Related to the very local loading, the carbon fibres store only a small amount of elastic energy. After sudden failure of the carbon fibres, the steel fibres continue to yield without failure and sustain further deflection. The enhanced maximum deflection and the load transfer to adjacent regions leads to an increase in energy absorption of approx. 25 % for SCFRP 10 and more than 50 % for SCFRP 20 in comparison with pure CFRP, Fig. 9.17. Positioning the steel fibres in the top or core layers of the laminate can further enhance the damage tolerance of the composite. In case of SCFRP 20i, the stiff and strong carbon fibres in the top layer enable a higher resistance against bending, while the steel fibres in the core layer can plasticise and maximise the structural integrity. The energy absorption can be increased by 145 %. In case of SCFRP 20a the steel fibres in the outer plies sustain the bearing cross section until higher deflections. By this means, the energy absorption can be improved by more than 170 %. The experimental findings of the electrical measurements show a pronounced increase of electrical conductivity as a function of stainless steel fibres volume content. The hybrid composite reinforced with 20 vol% stainless steel fibres (SCFRP 20) are characterised by the highest electrical conductivity, which is about five times higher than conventional CFRP. The observed increased standard deviation of the SCFRP in comparison to CFRP is explainable by the complex contact conditions due to the three different phases of the SCFRP and the copper electrodes. An influence of the hybrid concept (separated or homogeneous layer approach) could not be observed. However, the full theoretical potential could not yet be verified by these experiments, as the measured conductivities remained below the calculated values. Further investigations are necessary in order to understand the root causes.

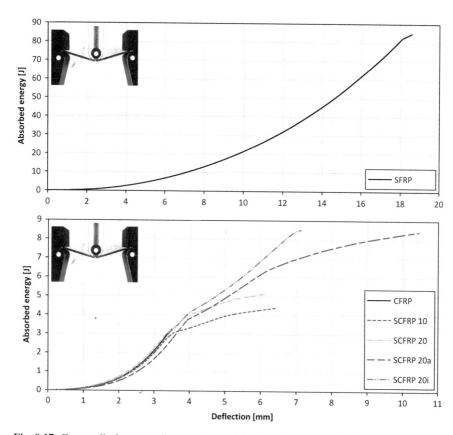

Fig. 9.17 Energy-displacement diagrams for combined tension-bending load

Conclusion and Outlook

Compared to state of the art CFRP, significant enhancements of electrical and mechanical key properties could be demonstrated by incorporation of stainless steel fibres into CFRP. The electrical conductivity can be increased. Also the potential for energy absorption and structural integrity can significantly be improved. The results suggest that an allocation of steel fibres in core or top layers is advantageous compared to a homogenous steel fibre distribution within the laminate. However, in case of pure tensile load, the deformability of the stainless steel fibre could not yet be fully exploited to obtain a global post failure process. Future steps will address the available deformability of the incorporated steel fibres. Lateral to the fibre orientation, the integrated steel fibres increase the stiffness of the composite, but lower its strength. Future work will therefore focus on complex, multiaxial laminates with (electrical and mechanical) optimised lay-up, complemented by further analysis, e.g. determination of compression after impact strength (CAI), penetration resistance, bearing strength, fatigue, corrosion resistance. Furthermore, the usability of integrated austenitic steel fibres for in-situ health monitoring (based on phase transition and the change of electrical resistivity) and inductive heating for advanced manufacturing processes will be investigated.

Shape Memory Alloy Wires: A New Approach to Morphing Structures

Moritz Hübler and Ulf Breuer

During the last century, the required structural and system mass to transport one passenger over a range of 1000 km was reduced by a factor of approx. 10, as shown in Fig. 9.18. However, during the last two decades, only a small reduction was archived. A reason for this rather limited reduction is the increased functionality of the airframe, especially in terms of safety and comfort. Additional functional elements required for safety and comfort add mass and compromise the mass reduction achieved by modern airframe design. However, the extensive use of CFRP in the latest aircraft (Boeing 787, Airbus A350) enabled a further improvement.

As the share of CFRP in the structural mass already reached more than 50 %, a further significant weight reduction requires fundamentally new solutions.

Improvements can be expected from:

- new high performance materials with higher strength, stiffness and improved damage tolerance (see Chap. 3)
- advanced manufacturing technology (see Chap. 4)
- advanced design principles (see Chap. 7)
- tailored structures with load-deflection coupling (see Chap. 8)
- lightweight actuators and systems [30, 31]
- function integration

Function Integration

A promising approach towards further significant weight savings is the function integration or the use of multifunctional components. The diversity of requirements and opportunities in Fig. 9.19 illustrates some examples of this approach to improve the complete aircraft system.

In general, compared to conventional solutions, a multifunctional part should demonstrate an improved relationship of operator benefits to manufacturing cost. Reduced space requirements and less weight can be important contributors. New functions which cannot be provided by conventional solutions can mean added product value, too.

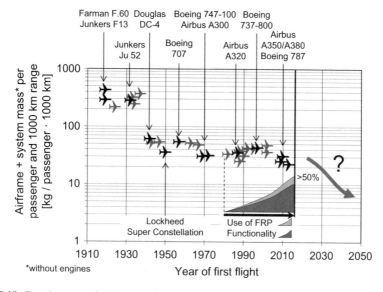

Fig. 9.18 Development of airframe and system mass during the last century

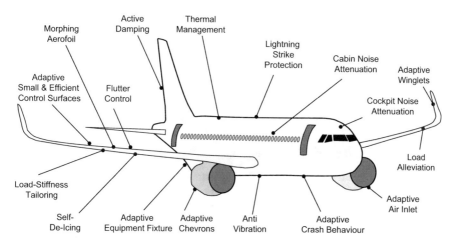

Fig. 9.19 Potential applications and opportunities for multifunctional materials

However, drawbacks of multifunctional parts can be linked to a more complex design process, and to precautions which can be necessary to compensate the reduced robustness if two or more functions run on one single part [32].

There are several examples in the common airframe structure, where function integration is already used. For example, the wing shells, ribs and spars bear the aerodynamic load and at the same time provide fuel storage. An aluminium or Glare® fuselage structure fulfils structural as well as electrical requirements. Unfortunately, this is not the case for a state-of-the-art CFRP fuselage, Fig. 9.20.

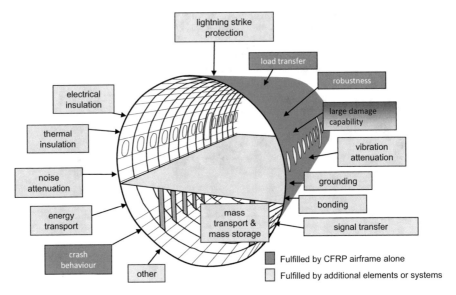

Fig. 9.20 CFRP fuselage functions (selection)

Except from part assembly, multi-functionality is strongly related to material properties. Functions already covered by common CFRP materials are load transfer, robustness, crash behaviour, etc. Other important functions not yet fully covered by composite materials are:

- electrical conductivity
- electromagnetic interference shielding (EMI)
- sensing and actuation
- active damping
- energy harvesting, energy storage

Tailored material properties can be transferred into application related functions, e.g. high electrical conductivity for grounding or lightning strike protection, or low thermal conductivity for the thermal insulation of the cabin. Sensing capabilities could increase the functionality of the airframe by enabling superior structure health monitoring (SHM) systems. A material integrated actuation, enabling fully controlled morphing structures, would offer completely new possibilities to optimise aerodynamics [33, 34].

As a simple example, Fig. 9.21 illustrates the situation for a separated and an integrated solution for the following functions: To provide a certain structure stiffness for a given surface and to enable a defined movement of this structure. While for the function separated solution an additional mechanical system is necessary for the actuation, the multifunctional material fulfils both functions.

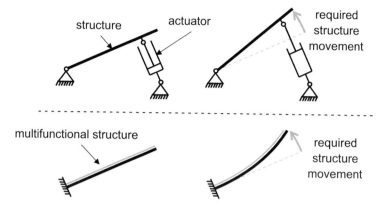

Fig. 9.21 Structure actuation by a conventional separated (*top*) and an integrated actuation (*bottom*) approach

Solid State Actuators

Multifunctional materials can convert an energy input of one type into an output of another energy type. In case of solid state actuator materials, different input energies (e.g. electric, magnetic, thermal) can be converted in a mechanical output, usable for actuation purposes [30]. Solid state actuators with their material-based actuation principle are predestined to be integrated in differential structure of fibre reinforced plastics FRP.

To compare the performance of available solid state actuators, the typical stress and strain values are illustrated in Fig. 9.22. The product of stress and strain defines the volumetric energy density, which indicates the required volume or design space. The solid state actuators differ strongly in the maximum possible actuation frequency for oscillating movements.

The most common solid state actuators are shape memory alloys (SMA) and piezoceramics. While piezoceramics are limited in terms of stress and strain, their dynamic response is outstanding. Combined with a reverse sensing capability, a broad range of applications is possible, e.g. noise attenuation, damping, structure health monitoring, actuation with small deflections, acoustic emission etc. SMA can deliver high performance actuation with strain values up to approx. 6 % and stress values up to approx. 600 MPa. However, due to the temperature driven effect, the dynamic of the reverse actuation is limited [30].

A comparison made by The Boeing Company between a SMA torque actuator of 0.5 kg delivering 17 Nm and an equivalent common solution with gear box and motor weighing 18 kg demonstrates the benefits of SMA actuators for applications with high forces and low duty cycles, in case weight is premium [36].

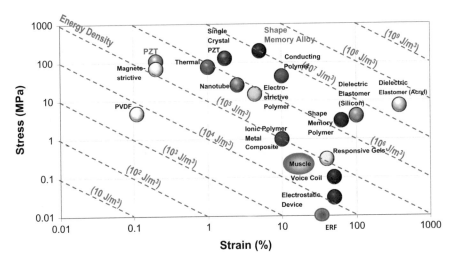

Fig. 9.22 Actuation performance of available solid state actuators according to [35]

Shape Memory Alloys

The memory effect has its origin in a thermos elastic phase transformation between the preferred configurations of the metal lattice. Shape memory alloys e.g. NiTi alloys have a high temperature phase, the austenite, and a low temperature phase, the martensite.

Figure 9.23a explains the basic one-way effect, starting with the deformation of the martensite by detwinning the natural twin structure. Due to this detwinning, a major part of the deformation remains in the material after unloading. A subsequent heating initiates the transition to austenite and the material is forced to take its original shape, as the austenite needs to adopt its only available lattice structure. This deformation is the so-called "memory effect" and can be used for actuation purposes. During load free cooling the transition to twinned martensite takes place. The actuation deformation can be adjusted by a pre-deformation of the material.

The two-way effect shown in Fig. 9.23b is a special case of the mechanism described before. By introducing dislocations and defects during a thermo mechanical training process, a partially detwinned martensite receives the preferred configuration of the material at low temperatures. By heating and cooling, the material changes between to shapes.

For the practical applications different alloy compositions with transition temperatures from 20 to 120 °C can be chosen. In some use cases, the ambient heat generated for example by the failure of a structural part or by an engine can be used to trigger the actuation. For a well-controlled actuation on demand, the joule heating due to an electric current by applying an electric voltage is an appropriate way to initiate the transition [30, 37].

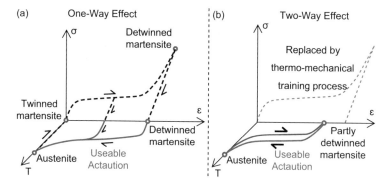

Fig. 9.23 One-way effect and two-way effect of shape memory alloys

Active Shape Memory Alloy: Fibre Reinforced Plastics Hybrid Structures

A possible method for active SMA FRP hybrid structure design is proposed in Fig. 9.24. After the assessment of suitable manufacturing process parameters (time, temperature, pressure) it is necessary to define the integration method, i.e. the SMA and laminate architecture.

The integrated SMA can be modelled as an actuator acting against a certain structure stiffness. It is important to assess the actuation characteristics of the chosen SMA alloy by means of experiments. Finite element analysis is possible by defining shell elements. The hybrid structure behaviour can be characterised on coupon level, and FEA can be applied to calculate the deflection of components. Finally, this calculation must be verified by experimental component testing.

SMA FRP Manufacturing and Integration

Different SMA alloys are commercially available in the form of wires. Diameters range from approx. 30 μm to 3 mm. In principal, SMA wires can be integrated in FRP by means of existing manufacturing concepts (for example resin infusion technology, see Chap. 3). However, for a direct integration of SMA into FRP the main challenges are:

- identification of suitable process temperatures during manufacturing of the integrated SMA-FRP component
- enabling a sufficient load transfer between SMA and FRP during operation

During the manufacturing of active hybrid SMA-FRP structures, the manufacturing process temperatures have to be adjusted to the polymer matrix (depending on viscosity and curing characteristics) *and* to the thermal behaviour of

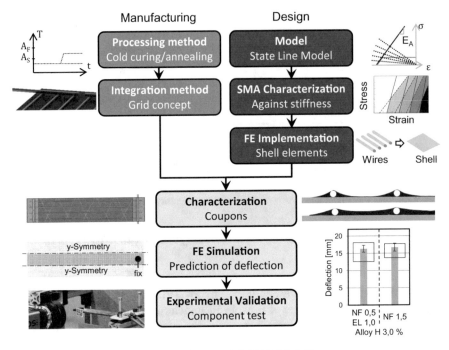

Fig. 9.24 Holistic design approach for integrated SMA FRP hybrid structures

the active SMA. While matrix polymers with superior mechanical performance at elevated operating temperatures usually also require elevated process temperatures, is it necessary to prevent any unwanted actuation of the SMA during manufacturing, see Fig. 9.25.

The attempt to mechanically suppress the SMA actuation by clamping means additional effort and can lead to poor reproducibility [38]. Curing at relatively low temperatures (below A_S) to consolidate the structure, followed by annealing, can lead to better results. Ideally, the thermoset polymer should provide sufficient adhesion between the SMA and the adjacent FRP already during the first curing cycle at lower temperatures (below A_S), and enable a post-curing (annealing) process, in which the SMA actuation potential is not diminished by any unwanted relative movement between the SMA and the FRP, finally leading to the physical and chemical properties needed for the practical application.

Superior mechanical properties of the cured matrix and sufficient adhesion between SMA and FRP are important to ensure the load transfer between the SMA elements and the FRP during actuation. As Fig. 9.26 illustrates, a direct integration of SMA in FRP within a beam (see also Fig. 9.25) does not lead to a homogeneous stress distribution on the SMA-FRP interface.

By embedding SMA wires or fibres within a bending beam, a high shear stress appears within the SMA-FRP interface at the ends of the beam if the SMA is

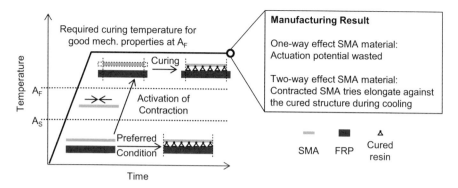

Fig. 9.25 Contradiction between elevated FRP process temperatures and SMA actuation

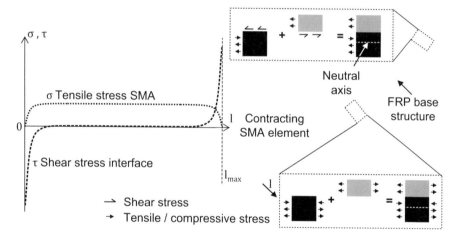

Fig. 9.26 Distribution of tensile stress within in the SMA element and SMA-FRP interface shear stress

activated, Fig. 9.26. Without any special load introduction concept, interface failure close to the stress peaks is very difficult to avoid in practise.

The concept shown in Fig. 9.27 is based on an integrated semi-finished SMA grid, which provides a mechanical interlocking by means of an "anchor" wire, transferring load between the activated SMA wires and the surrounding FRP. This anchor area can be reinforced by additional FRP with continuous or discontinuous fibres. The anchor wire itself—although consisting of the same SMA alloy as the active wires—is inactive during joule heating or cooling.

The approach to use a semi-finished SMA wire grid enables simple handling and precise positioning during the manufacturing process. The grid can also be used to generate an electrical network for activation on demand by joule heating. Tailoring the grid with respect to wire diameter, wire spacing, wire length etc., is also possible.

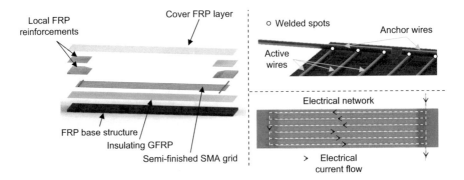

Fig. 9.27 Integration of an SMA grid for mechanical interlocking by anchor wires

Shape Memory Alloy Characterisation and Finite Element Model

In order to follow an engineering-friendly design method for morphing structures based on this SMA-FRP concept as proposed in Fig. 9.24, it is first of all necessary to characterise the actuation behaviour of the active SMA material while working against a certain spring stiffness. Similar to a spring, the deformed FRP component will show an increased counterforce with increasing deflection.

An experimental setup as well as some typical characterisation results for a SMA material are presented in Fig. 9.28. Additional details can be found in [39].

These results can then be used for an FEA based design approach as described in [39] and [35]. Figure 9.29 illustrates the principle of transferring the micro-section to a simple shell description, which can be applied for FEA.

With a simplified model of state lines, where each state line represents a certain temperature, internal transformations within the SMA material can be taken into account by introducing a temperature dependent strain term, Fig. 9.30. A single state line can be described by the stiffness and the free strain. The implementation of an additional strain term ε_{SMA}, similar to a linear temperature expansion, can be used to model the actuation. The total strain ε can be calculated by adding the strain according to Hooks law (σ/E_{SMA}) and the actuation strain ε_{SMA}. The latter is calculated by the product of the strain coefficient β_{SMA} and the variable χ_{SMA} (for a certain temperature). χ_{SMA} ranges from 0 % (actuation "off") to 100 % (actuation "on"). An excellent description of the principle of the FEA implementation, a proposal for the assessment of β_{SMA} and χ_{SMA} as well as examples can be found in [35]. Also a self-sensing possibility is described.

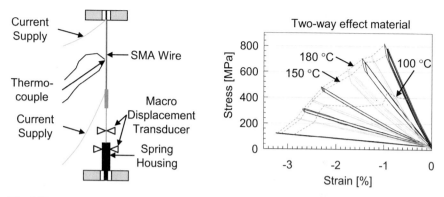

Fig. 9.28 Characterisation of SMA actuation behaviour working against different spring stiffness [39]

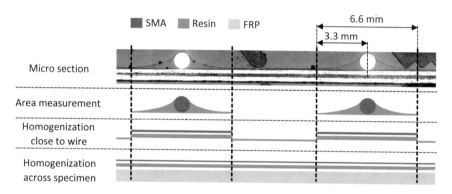

Fig. 9.29 From micro geometry to a shell description of the active structure [39]

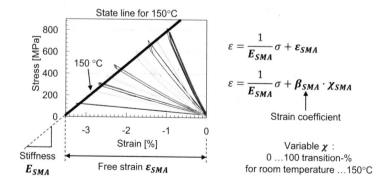

$$\varepsilon = \frac{1}{E_{SMA}}\sigma + \varepsilon_{SMA}$$

$$\varepsilon = \frac{1}{E_{SMA}}\sigma + \beta_{SMA} \cdot \chi_{SMA}$$

Strain coefficient

Variable χ :
0 …100 transition-%
for room temperature …150°C

Fig. 9.30 Description of an exemplary state line and the calculation of the total strain [39]

Potential New Applications

Adaptive Winglets

For commercial aircraft and business jets, there is wide range of different wing tip designs. In general, the attempt is to reduce the induced drag of the vortex generated at the wing tip. However, wing tips are additional elements, and their efficiency depends on the air speed. To provide the best performance also for varying cruise speed and for different flight phases, a situation dependent shape adjustment of the winglets would be beneficial.

However, the design space for actuator systems is strongly limited at the wing tip, and any additional mass also has a strong impact on the dynamic behaviour. Morphing structures made by multifunctional material can enable advanced solutions. The benefit has already been verified by means of wind tunnel tests [40] and patents filed by The Boeing Company [41].

Adaptive Vortex Generator

High performance aerofoil can show minimum drag and maximum lift, but they can also tend to sudden stall due to flow separation at very low air speed. Increased approach speed can thus be necessary, resulting in less ability for steep descend, and potentially in a higher noise foot print. For some aircraft, this problem is solved with static vortex generators on the wing surface, which prevent flow separation by energising the boundary layer of air flow. However, this concept causes permanently increased drag during cruise. Active vortex generators, deployed only on demand at low speed, can help to overcome this problem.

As schematically shown in Fig. 9.31 for a glider, a number of vortex generators distributed over the wing surface can be necessary. Instead of a space-consuming central actuator, connected to each individual vortex generator by complex mechanisms, an independent actuation powering each vortex generator element can be beneficial. Mechanisms partly driven by SMA actuators have been investigated [42–44]. However, it is a challenge to ensure a completely smooth surface for the retracted position.

Adaptive Air Inlet Lip

Reducing the drag to a minimum is another target of adaptive air inlets lips investigated for military jet engines. The size of air inlet of the engine is adjusted to the current demand of the turbine, allowing the engine to operate with higher efficiency. Two actuators with embedded SMA rods in an elastomeric matrix combined in a protagonist antagonist system where used in prototype tests, allowing a controlled two-way actuation and positioning without energy demand [45].

Fig. 9.31 Situational adaption of aerodynamics by adaptive vortex generators

Adaptive Chevrons

Chevrons guide the air flow at the turbine outlet, influencing efficiency and noise. Due to densely populated airport neighbourhoods and an increasing traffic, noise reduction systems are of high interest. Adaptive chevrons help to reduce engine noise on demand during ground level flight manoeuvres, while the efficiency of the engine especially during cruise can be maximised. State of the art chevrons are thin metal extensions of the turbine housing. The design space is very limited. For the development of active chevron prototypes, three shape memory alloy bending actuators per chevron can be mechanically integrated in order to reduce the section of the engine outlet on demand [46–48].

Adaptive Trailing Edge

Small adjustments of the trailing edge of the wing can have a strong effect on the aerodynamic performance. A trailing edge mini flap can allow high climb rates, reducing the time close to populated airports and noise impairment. A mini flap has to be powered by a lightweight actuator with minimum size. Extensive research work has been performed at Airbus [49, 50]. Investigations on this topic applying a SMA torsion actuator were also performed by The Boeing Company and flight tested [51].

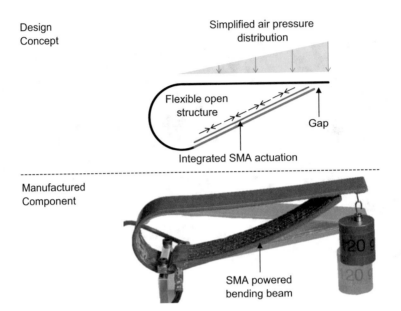

Design
Concept

Simplified air pressure
distribution

Flexible open
structure

Gap

Integrated SMA actuation

Manufactured
Component

SMA powered
bending beam

Fig. 9.32 Active aerofoil demonstrator

Slat-Cove Filler

One of the potential sources of noise is the cove behind a deployed slat. By filling this cove, the generation of a vortex can be suppressed and the noise exposure is significantly reduced. Passive elements made of pseudo-elastic SMA can be applied to fill the cove, and the movement of the slat can trigger their deployment and retraction. Instead of the actuation potential of SMA, their pseudo-elastic behaviour is used for a reliable deployment of the cove filler. The pseudo-elastic effect is available if the material in its austenite phase is forced to transform into a detwinned martensite by applied stress. This reversible transformation allows high strain at a constant stress level [52].

Active Aerodynamic Aerofoil

By integrating thin SMA filaments unsymmetrically within a fibre reinforced polymer laminate (see Fig. 9.27) it becomes possible to engineer a multifunctional bending beam, Fig. 9.21. This material integration can result in cost and weight advantages when compared to common structure-actuator combination and to differential structure-SMA actuator combinations:

- no mechanics and couplings are required
- the number of parts is reduced
- very low installation space is required
- the actuator output is not applied to discrete points of the structure
- the actuator is continuously scalable

To investigate these advantages, an active aerodynamic aerofoil was developed by means of a SMA powered bending beam. The principle of the solution can also be applied for other applications such as active wing trailing edges. A first technology demonstrator is shown in Fig. 9.32.

The active aerofoil demonstrator verified the actuation performance of control surfaces of gliders. Both torque and actuation angle can be further increased by adjusting temperature, the amount of SMA integrated within the composite laminate and the layup configuration [39].

References

1. Gillot, A.: From carbon fibre to carbon-fibre-reinforced thermoplastics. JEC Composites Magazine, no. 71 (2012)
2. The Outlook for thermoplastics in Aerospace Composites – 2014–2023. Information on http://www.compositesworld.com/ (2014)
3. Carbon fiber, PEEK combined in lightweight aerospace composite. Information on http://www.plasticstoday.com/ (2013)
4. PEEK resin offers lightweighting option for Airbus. Information on http://www.plasticstoday.com/ (2013)
5. Paul, D.R., Bucknall, C.B.: Polymer Blends. Formulation, vol. 1. Wiley, New York, NY (2000)
6. Salem, D.R.: Structure Formation in Polymeric Fibers. Carl Hanser Verlag, München (2001)
7. Macosko, C.W.: Morphology development and control in immiscible polymer blends. Macromol. Symp. **149**, 171–184 (2000)
8. Nwabunma, D., Kyu, T.: Polyolefin Blends. Wiley, Hoboken, NJ (2008)
9. Robeson, L.M.: Polymer Blends, A Comprehensive Review. Carl Hanser Verlag, München (2007)
10. Carothers, W.H., Hill, J.W.: Studies of polymerization and ring formation: xv. Artificial fibers from synthetic linear condensation superpolymers. J. Am. Chem. Soc. **54**, 1579–1587 (1932)
11. Grace, H.P.: Dispersion phenomena in high viscosity immiscible fluid systems and application of static as dispersion devices in such systems. Chem. Eng. Commun. **14**, 225–277 (1982)
12. Favis, B.D., Chalifoux, J.P.: The effect of viscosity ratio on the morphology of polypropylene/polycarbonate blends during processing. Polym. Eng. Sci. **27**(20), 1591–1600 (1987)
13. Krooß, T., Gurka, M., Dück, V., Breuer, U.: Development of cost-effective thermoplastic composites for advanced airframe structures. ICCM20, Copenhagen (2015)
14. Lai, M., Liu, J.: Thermal and dynamic mechanical properties of PES/PPS blends. J. Ther. Anal. Calorim. **77**, 935–945 (2004)
15. Shibata, M., Yosomiya, R., Jiang, Z., Yang, Z., Wang, G., Ma, R., Wu, Z.: Crystallization and melting behavior of poly(p-phenylene sulfide) in blends with poly(ether sulfone). J. Appl. Polym. Sci. **74**, 1686–1692 (1999)
16. Karam, H.J., Bellinger, J.C.: Deformation and breakup of liquid droplets in a simple shear field. Ind. Eng. Chem. Fund. **7**(4), 576–581 (1968)
17. Tavgac, T.: Ph.D. thesis, Houston, Texas (1972)
18. Auer, C., Kalinka, G., Krause, T., Hinrichsen, G.: J. Appl. Polym. Sci. **51**, 407–413 (1994)
19. Spruiell, J.E.: A review of the measurement and development of crystallinity and its relation to properties in neat poly(phenylene sulfide) and its fiber reinforced composites. Technical Report. (2005). doi:10.2172/885940
20. Nishihata, N., Koizumi, T., Ichikawa, Y., Katto, T.: Plane strain fracture toughness of polyphenylene sulfide. Polym. Eng. Sci. **38**(3), 403–408 (1998). doi:10.1002/pen.10201

21. Bold, J.: Airbus composites training: skills for composites. MSC Software Virtual Product Development Conference, Frankfurt, 2007
22. Hellard, G.: Composites in Airbus. A Long Story of Innovations and Experiences. EADS Global Investor Forum, Sevilla, 2008
23. Breuer, U., Schmeer, S., Eberth, U.: Carbon and metal fibre reinforced airframe structures – a new approach to composite multifunctionality. Deutscher Luft- und Raumfahrtkongress 2013, Stuttgart, 10–12 Sept 2013
24. Spitalsky, Z., Tasis, D., Papagelis, K., Galiotis, C.: Carbon nanotube – polymer composites: chemistry, processing, mechanical and electrical properties. Prog. Polym. Sci. **35**(3), 357–401 (2010)
25. Callens, M.G., Gorbatikh, L., Verpoest, I.: Ductile steel fibre composites with brittle and ductile matrices. Compos. A: Appl. Sci. Manuf. **61**, 235–244 (2014). doi:10.1016/j.compositesa.2014.02.006
26. Radtke, A., Van Wassenhove, V., Van Koert, K., Vöge, F., Moors, O., Fertig, D.: Reinforcing with steel cord. In: Kunststoffe International. Hanser Verlag, München (2012)
27. Schmeer, S., Steeg, M., Maier, M., Mitschang, P.: Metal fiber reinforced composite – potentialities and tasks. Adv. Compos. Lett. **18**(2) (2009), http://www.acletters.org/abstracts/18_2_2.html
28. Vandewalle, S.: Multifunctional steel fibre reinforcement. Wolfsburg Innovations in Automotive Textiles. 07.10.2010.
29. Toho Tenax Europe GmbH: HTS Delivery program and characteristics, Version 05, 2014-09-30. http://www.tohotenax-eu.com. Reviewed: 15 May 2015
30. Janocha, H.: Adaptronics and Smart Structures, Basics, Materials, Design and Applications, 2nd revised edn. Springer, Berlin (2007). ISBN 978-3-540-71965-6
31. Gloess, R., Lula, B.: Challenges of extreme load hexapod design and modularization for large ground-based telescopes. Proceedings of SPIE – The International Society for Optical Engineering, July 2010
32. Ehrlenspiel, K.: Integrierte Produktentwicklung. Carl Hanser Verlag, München (2009)
33. Hübler, M., Gurka, M., Schmeer, S., Breuer, U.: Performance range of SMA actuator wires and SMA–FRP structure in terms of manufacturing, modeling and actuation. Smart Mater. Struct. **22** (2013). doi:10.1088/0964-1726/22/9/094002
34. Hübler, M., Nissle, S., Gurka, M., Breuer, U.: Load-conforming design and manufacturing of active hybrid fiber reinforced polymer structure with integrated shape memory alloy wires for actuation purposes. Proceedings, ACTUATOR 2014, 14th International Conference on New Actuators, Bremen, 23.-25. Juni 2014
35. Gurka, M.: Active hybrid structures made of shape memory alloys and fibre-reinforced composites. In: Friedrich, K., Breuer, U. (eds.) Multifunctionality of Polymer Composites, pp. 727–751. Elsevier, Philadelphia, PA (2015). ISBN 978-0-323-26434-1
36. Urnes, J., Nguyen, N., Dykman, J.: Development of Variable Camber Continuous Trailing Edge Flap System. Fundamental Aeronautics Technical Conference, 13 Mar 2012
37. Lagoudas, D.C.: Shape Memory Alloys. Springer, New York, NY (2008)
38. Hübler, M., Gurka, M., Breuer, U.P.: From attached shape memory alloy wires to integrated active elements, a small step? Impact of local effects on direct integration in fiber reinforced plastics. J. Compos. Mater. **49**(15), 1895–1905 (2015)
39. Hübler, M.: Methodik zur Auslegung und Herstellung von aktiven SMA-FKV-Hybridverbunden. Schriftenreihe Institut für Verbundwerkstoffe GmbH, Kaiserslautern (2015)
40. Srikanth, G., Surendra, B.: Experimental investigation on the effect of multi-winglets. Int. J. Mech. Indus. Eng. **1**(1), 43–46 (2011)
41. Pub. No.: US 2008/0308683 A1: CONTROLLABLE WINGLETS. THE BOEING COMPANY, Inventors: Mithra M.K.V. Sankrithi, Lake Forest Park, WA (US); Joshua B. Frommer, Seattle, WA (US); Pub. Date: Dec. 18, 2008

42. Quackenbush, T.R., McKillip, M.R., Whitehouse, G.R.: Development and testing of deployable vortex generators using SMA actuation. 28th AIAA Applied Aerodynamics Conference, Chicago, IL, 28 June–1 July 2010

43. Pub. No.: US 6,427,948 B1: CONTROLLABLE VORTEX GENERATOR. Inventor: Michael Campbell (US); Pub. Date: Aug. 6, 2002

44. Pub. No.: US 2014/0331665 A1: VORTEX GENERATOR USING SHAPE MEMORY ALLOYS. THE BOEING COMPANY, Inventor: Belur N. Shivashankara, James H. Mabe, Dan J. Clingman (US); Pub. Date: Now. 13, 2014

45. White, E.V.: Defense system perspectives on multifunctional design for actuation. The 2nd Multifunctional Materials for Defense Workshop, 30 July–3 Aug 2012

46. Hartl D.J., Mooney, J.T., Lagoudas, D.C., Calkins, F.T., Mabe, J.H.: Use of a Ni60Ti shape memory alloy for active jet engine chevron application: II. Experimentally validated numerical analysis. Smart Mater. Struct. **19**(1) (2009)

47. Hartl D.J., Mooney, J.T., Lagoudas, D.C., Calkins, F.T., Mabe, J.H.: Use of a Ni60Ti shape memory alloy for active jet engine chevron application: II. Experimentally validated numerical analysis. Smart Mater. Struct. **19** (2010)

48. Pub. No.: US 2009/0301094 A1: GAS TURBINE ENGINE EXHAUST NOZZLE HAVING A COMPOSITE ARTICLE HAVING A SHAPE MEMORY MATERIAL MEMBER. ROLLS-ROYCE PLC, Inventor: John Richard Webster, Derby (GB); Pub. Date: Dec. 10, 2009

49. Breuer, U., Latrille, M.: Einsatzmöglichkeiten und Restriktionen adaptiver CFK-Strukturen bei Verkehrsflugzeugen. Proceedings Adaptronic Congress Potsdam, 4.-5.4.2000

50. Breuer, U., Jänker, P., Lorkowski, T.: Linear, hydraulic pivot drive. US Patent 7028602

51. Wilsey, C.: Continuous Lower Energy, Emissions and Noise (CLEEN) Technologies Development Boeing Program Update. LEEN Consortium Public Session, 2012

52. Turner, T.L., Kidd, R.T., Hartl, D.J., Scholten, W.D.: Development of a sma-based, slat-cove filler for reduction of aeroacoustic noise associated with transport-class aircraft wings. Proceedings of the ASME 2013 Conference on Smart Materials, Adaptive Structures and Intelligent Systems SMASIS2013, Snowbird, UT, 16–18 Sept 2013

53. Noll, A.: Effektive Multifunktionalität von monomodal, bimodal und multimodal mit Kohlenstoff-Nanoröhrchen, Graphit und kurzen Kohlenststofffasern gefülltem Polyphenylensulfid, vol. 98. IVW Schriftenreihe, Kaiserslautern (2012)

Further Reading

1. Garg, C.A., Mai, Y.W.: Failure mechanisms in toughened epoxy resins – a review. Compos. Sci. Technol. **31**(3), 179–223 (1988)

2. Medina Barron, R.M.: Rubber Toughened and Nanoparticle Reinforced Epoxy Composites, vol. 84. IVW Schriftenreihe, Kaiserslautern (2009)

3. Gibson, R.F.: A review of recent research on mechanics of multifunctional composite materials and structures. Compos. Struct. **12**, 2793–2810 (2010)

4. Wulz, H.G., Petricevic, R., Gurka, M.: New Advanced Materials Smart Materials. ECSS E-30-04 ESA Structural Material Handbook ESA ESTEC, Nordvijk, 2006

5. Laufenberg, M.: Magnetische Formgedächtnisaktorik: Die bessere Alternative, VDI-FGL Expertenforum, Dresden, 11.06.2013

Index

© Springer International Publishing Switzerland 2016
U.P. Breuer, *Commercial Aircraft Composite Technology*,
DOI 10.1007/978-3-319-31918-6